T0155761

# Synthese Library

Studies in Epistemology, Logic, Methodology, and Philosophy of Science

Volume 410

The aim of *Synthese Library* is to provide a forum for the best current work in the methodology and philosophy of science and in epistemology. A wide variety of different approaches have traditionally been represented in the Library, and every effort is made to maintain this variety, not for its own sake, but because we believe that there are many fruitful and illuminating approaches to the philosophy of science and related disciplines.

Special attention is paid to methodological studies which illustrate the interplay of empirical and philosophical viewpoints and to contributions to the formal (logical, set-theoretical, mathematical, information-theoretical, decision-theoretical, etc.) methodology of empirical sciences. Likewise, the applications of logical methods to epistemology as well as philosophically and methodologically relevant studies in logic are strongly encouraged. The emphasis on logic will be tempered by interest in the psychological, historical, and sociological aspects of science.

Besides monographs *Synthese Library* publishes thematically unified anthologies and edited volumes with a well-defined topical focus inside the aim and scope of the book series. The contributions in the volumes are expected to be focused and structurally organized in accordance with the central theme(s), and should be tied together by an extensive editorial introduction or set of introductions if the volume is divided into parts. An extensive bibliography and index are mandatory.

More information about this series at http://www.springer.com/series/6607

Arnold Koslow

# Laws and Explanations; Theories and Modal Possibilities

Arnold Koslow
Department of Philosophy, The Graduate Center
City University of New York
New York, NY, USA

Synthese Library
ISBN 978-3-030-18848-1    ISBN 978-3-030-18846-7 (eBook)
https://doi.org/10.1007/978-3-030-18846-7



This Springer imprint is published by the registered company Springer Nature Switzerland AG.
The registered company address is: Gewerbestrasse 11, 6330 Cham, Switzerland

*To Julian and Jennifer*
*For their love, and support,*
*and in memory of*
*Sidney Morgenbesser and Isaac Levi*
*Amor mi mosse, che mi fa parale*

# Contents

# Part I
# Burnishing the Legacy

# Chapter 1
# Introduction

This essay on scientific laws has two parts. (I) **Part I**, (Chaps. 2, 3, 4, 5, 6 and 7), (Burnishing the Legacy, *Laws and explanations.*) and **Part II** (Chaps. 8, 9, 10, 11 and 12), Devoted to the explanation of laws by theories: *Schematic and non-schematic theories, two kinds of explanation, and the physical modals that each theory provides, and the nomic modals that are associated with laws.*

Part I: Chapter 2, begins with a discussion of the classical paper of Hempel and Oppenheim, and focuses on the account of laws in that essay. Chapter 3 "Laws and their corresponding counterfactuals; an untenable connection" then considers the possibility of saving the Hempel Model from some telling criticism of it, by invoking counterfactuals. Chapter 4 examines the influential and almost total rejection of the Hempelian account of laws by F. Dretske and the final Chaps. 5 and 6 of the first part are devoted to the well known proposal of D. Armstrong whose proposed account of laws rested on the exclusive reliance on his theory of universals.

The emphasis in the six chapters is not to provide new counterexamples for old theories. There's been enough of that already. The stress is on burnishing those classic accounts. Counterexamples tell you that the accounts are defective, but that's it; and usually all of it. We want to look at the best versions of those accounts, to learn from them by perhaps supplementing them with some friendly amendments – even though, in the end, they may, and do fall short. So I think of the first five chapters as a burnishing of those old theories; not merely a berating of them.

**Chapter 2**. The proposal by Hempel and Oppenheim has to do more with explanation, than with laws. The structure they provided for explanation was deductive (we omit for the present their discussion of probabilistic explanation), with laws being characterized as a kind of general sentence that is essential for the deduction. Hempel chose a universally quantified material conditional for the representation of laws. That choice however was the implicationally weakest sentence to fill the inferential role that laws were supposed to have. Hempel could have deflected a lot of criticism had he taken laws to be the universally quantified counterfactual conditionals. He didn't. It is worth noting that both Dretske, and later, Armstrong made sure that laws on their accounts were implicationally stronger



A. Koslow, *Laws and Explanations: Theories and Modal Possibilities*, Synthese Library 410, https://doi.org/10.1007/978-3-030-18846-7_1

than universally quantified material conditionals. There is of course a certain irony that in choosing implicationally stronger statements for laws, Dretske and Armstrong were satisfying the Hempelian project, charitably conceived.

Going counterfactual, however, wouldn't have saved the day because conditional laws are regarded as logically equivalent to their contrapositives. That is not true for counterfactuals. So representing laws as counterfactuals runs up directly against familiar scientific practice. That is the point of the third chapter.

In **Chapter 3** considers the familiar claim that laws support their corresponding counterfactuals. This thesis was usually used to define the difference between those generalizations that are laws from those generalizations that are accidental generalizations. That is neat but not true. We note is that there is a subtle ambiguity in the expression of the counterfactual connection

(C) Laws imply their corresponding counterfactuals (the *counterfactual connection*).

The problem is that there are at least two possible readings of the connection (C)

(1) If A is a law, then A implies its corresponding counterfactual conditional.
(2) $\mathcal{L}$(A) implies the counterfactual corresponding to A, where "$\mathcal{L}$(A)" states that it is a law that A.

The first says that if A is a law, then it is A that implies the corresponding counterfactual, while the second requires that the corresponding counterfactual follow from the statement "A is a law". There are interesting consequences of the adoption of either version. (1) has the unfortunate consequence that laws are equivalent to counterfactual conditionals, – something we argued against in the preceding chapter. (2), under plausible assumptions about corresponding conditionals, yields the nice conclusion that $\mathcal{L}$(A) implies A – that is, all laws are true.

**Chapter 4** considers F. Dretske's dramatic rejection of the Hempelian model. He drew a stark contrast with the standard model by a vigorous defense of the following four theses:

(1) Laws are *relations* between, properties, quantities, magnitudes, or features that are expressed by certain predicates (emphasis added).
(2) The relations in each law are extensional, and can vary from law to law,
(3) Laws are singular sentences rather than general, and
(4) The statement of the law All F's are G will not change it's truth-value if any co-referential expression replaces "F", "G", or both. However the result of that replacement may not be a law. That is, as he states, misleadingly, the opacity of laws.

This looks like a straightforward rejection of the central elements of the Hempelian account of laws. Dretske emphasized this difference by representing laws for example as (i) F-ness ➨ G-ness, (F-ness is related to G-ness) instead of the usual way, as a universally quantified material condition between predicates: (ii) $\forall$x (Fx $\supset$ Gx). In Dretske's representation the dark single-shafted arrow indicates a relation (not the connective for material conditional), and "F-ness", and "G-ness" stand for universals, not the predicates "Fx" and "Gx".

However, as we will show, one can minimize those differences by adopting a slightly amended account of universals that connects up those universals that occur in laws with certain corresponding predicates. The result is that the two representations (i) and (ii) are provably equivalent. (4) makes a simple error: In the expression "It is a law that A", you may replace predicates that occur in A by other predicates that are respectively coextensional so that the result of the substitution in A becomes (say) A∗. A and A∗ will have the same truth value, but A is a law and A∗ is not. This does not show (Dretske to the contrary) that A is opaque. It does show that "It is a law that A" is opaque, – another matter entirely ("opaqueness" standardly means that substitution of coextensional terms for some terms can result in a difference of truth-value.).

Dretske is only partly right about (2). Dretske's talk of relations between (say magnitudes) appeals to relations between universals, while Hempel and others talk about universal quantification of a material conditional between predicates. That is a different matter. The material conditional is a connective and certainly not a relation. Moreover Dretske asserts that the single shafted arrow (➨) differs from law to law, while the usual classical view has it that it is always the same material conditional connective. Even on his own view the laws he cites are usually represented as identities between the values of physical magnitudes. That relation doesn't change from law to law, as Dretske would have it.

The claim in (3), is that laws are not generalizations; they are singular statements, is just spectacular. Since on his account, laws are relations between singular intensional entities such as F-ness, and G-ness, or specific properties or magnitudes, he then claimed that they are, on those grounds, singular statements. Unfortunately there is a simple counter-example: Let R be the set of all ravens, and B the set of all black things. The sets are specific, and the set-theoretical relation of subset holds – i.e. R is a subset of B. But the law seems to be general, not singular: All ravens are black. I don't therefore put much stock in Dretske's way of placing his account in stark contrast to the usual view. It seems a bit of an exaggeration, and one should not make too much of it.

Thus given an elementary bit of a theory of universals, it would follow that Dretske's proposal that laws are singular statements involving extensional relations between intensional entities is not much of an advance over the received view of laws. It's not far off from the older view. In the end, we will show that his view is equivalent, given the supplement of an elementary theory of universals, to what the older Humean accounts maintained all along.

Chapters 5 and 6 are devoted to the proposal of D. Armstrong.

**Chapter 5**. Provides the mathematical and philosophical background for the detailed discussion of Armstrong's view in the following chapter. Chapter 5 begins with a discussion of the remarkable mid-Nineteenth century mathematical debate about whether the older prevalent intensional notion of a function should be replaced by an extensional account of functions that was no longer encumbered by the constraints imposed by the intensional accounts. The extensional view that eventually prevailed did so on the basis of an argument that maintained that the older intensional view was unable to express some scientific laws or even provide solutions to some outstanding scientific problems.

B. Russell (and A.N. Whitehead) kept to the older intensional notion when they deployed an intensional concept of propositional function in their *Principia Mathematica*. We believe that F. Ramsey was aware of the debate which raged on the continent, and he urged Russell to abandon his intensional notion of a propositional function for the second edition of the *Principia*. Ramsey argued that, hampered by their restriction to intensional functions, they could not define the identity relation, and consequently could not express basic truths of arithmetic, let alone provide proofs of them. I believe that Ramsey was inspired by the mathematical debate. Russell, and, as we shall see, Armstrong were left unmoved. Russell accepted Ramsey's criticism, but did not make the required changes for the second edition. I doubt whether Armstrong was aware of the mathematical debate on the continent, or of the exchange between Russell and Ramsey. Nevertheless, we shall argue in **Chap. 5**, that in the end, Armstrong's account of scientific laws is subject to a similar criticism. Despite the effort we shall make to supplement Armstrong's theory, it lacks critical expressive power: it doesn't have the means to express many of the laws that are the subject of his book.

**Chapter 6.** Armstrong's influential account of laws relies heavily on his account of universals., and as we shall argue, it is at a huge cost. Laws are described as second-order universals, N(F, G), which relate, say, the universals F, and G. The thought is, as he states it, that the universal F "necessitates" the universal G. Also in play are states of affairs (e.g. a particular a's instantiating a universal F, a particular b's instantiating the universal G, etc.), which we will denote by F[a], and G[b], etc. respectively. We are told that N(F,G) is a relation, but we are not told, nor is it assumed that it is a conditional. In addition, Armstrong assumes that there is a causal universal that holds or fails to hold between states of affairs. In addition, although there is no assumption that N(F,G) is a conditional, he assumes that N(F,G) implies a universally quantified material conditional between the predicates "Fx" and "Gx" which principle he calls *Modus Ponens* (**MP**), that is, $N(F, G) \Rightarrow (x)(Fx \supset Gx)$, but not conversely. This assumption implies that laws are at least as strong as their corresponding universally quantified material conditionals. This is an important assumption for Armstrong, but, unfortunately, it is not coherently expressed. There is no proof given for this implication, and it is clear that the left hand side of the implication contains references to the universals F and G, while the right-hand side contains references to certain predicates "Fx", and "Gx". Armstrong doesn't explain how the universals and predicates might be connected so as to insure that the implcation (**MP**) holds. We will rectify that gap with a friendly amendment that connects the instantiations of those universals that occur in the expression of laws with the instances of an associated predicate.

Armstrong is firmly convinced that the appeal to universals is a virtue, but it is far less than that, when it is the basis for discounting familiar laws like Newton's first law, that of inertia. That law in one formulation has a vacuous antecedent (If there are no forces on a body then . . .), but for Armstrong there are no universals that are empty) so it cannot be a law. On the other hand, the contrapositive formulation has an antecedent that is not empty, and is usually regarded as providing an equivalent formulation of the law of inertia. This is an obvious case of running headfirst against well entrenched scientific practice, and is a problem. Armstrong is aware of this, but it is still a problem.[1] There are more problems to come that are more serious still.

The central notion of N(F,G) is not explained, though it is sometimes illustrated, and the formal properties of N(F, G), are described, but not supported or proved. We will bolster Armstrong's account with two additions which we think are friendly ones. They are glosses drawn from various explanations that Armstrong has given.

The first explains N(F,G) as a universal generalization of a causal relation (R) between states of affairs F[x] and G[x] i.e.

(1) $N(F, G) \Leftrightarrow (x)R(F[x], G[x])$,

and the second condition requires that the statement that N(F, G) is a law, implies (x) R(F[x], G[x]). I.e.

(2) It is a law that $N(F, G) \Rightarrow (x)R(F[x], G[x])$.

With these friendly additions it can then be proved that all laws are true (Factivity holds for "It is a law that"), and one can now also provide proofs for the various formal conditions for N(F, G), that Armstrong described: *Irreflexivity*, *Non-Symmetry*, *Non-contrapositive*, and *Non-Transitive*, provided we make some reasonable assumptions about the causal relation that is supposed to hold between states of affairs.

Despite these amendments and additions, there remains a deep problem that I see no way of resolving. In Chap. 5 we discussed the flaw that F. Ramsey spotted in Russell and Whitehead's *Principia.* Essentially, the natural way of defining the identity relation for them was to use the notion of a propositional function which maps objects to propositions in such a way that if a is mapped to the proposition P(a), and object b is mapped to the proposition P(b), etc., then P(a) says of a, the same thing that P(b) says of b. The definition of the identity relation (a = b}) would be given as holding if and only if for every proposition P, P(a) if and only if P(b). The problem, as Ramsey noted, is that propositional functions so defined, are intensional functions, and one can show, that for that reason, it fails as a definition of identity.

Without an adequate concept of identity at hand, the account of arithmetic is expressively incomplete. That is a blow to the claim that arithmetic has been recovered, using the theory that Russell and Whitehead provided.

A similar fate we think befalls Armstrong's account. Following the pattern that Russell used, we can define state-of-affairs functions that for any universal U, the function PU assigns to each particular a the state of affairs U[a] (a's instantiating U), to particular b, the state of affairs U[b] (b's instantiating instantiating U), etc. etc. where U[a] says of a, the same thing that U[b] says of b. Thus this definition relies on what Armstrong says about the instantiation of universal by particulars. Roughly,

---

[1]In private conversation, Armstrong said that he hadn't thought of the contrapositive. I don't know what his solution might be, or if he returned to the issues involved. Clearly his distinction between derived and underived laws in *What is a Law of Nature*, will not help.

"It's the same damn thing again and again." Then we might try to define a = b as holding if and only if for every universal U, PU (a) if and only if PU (b). State-of affairs-functions are intensional and won't deliver an adequate notion of identity for the purposes that Armstrong needs – i.e. an identity relation that is needed for the expression not only of laws, but of physical magnitudes as well.[2]

This way of introducing identity follows the pattern of Russell's way of introducing it in the Principia. It didn't work there, and it doesn't work here. The fact is, however, that nothing would work for Armstrong. The simple reason, as he states explicitly, is that he believes that identity is not a relation. So there's no use looking for a technically suitable way of defining identity, nor would it help to just introduce it as a needed addition to his account. Armstrong, truth be told, would have none of it.

Our conclusion then, is that with the restriction to Armstrong's brand of universals, his account cannot even express most laws that need functions for their expression, and all physical magnitudes. That is a loss of the subject matter that the theory was supposed to discuss.

**Chapter 7.** Laws and Accidental Generalizations, A new minimal theory (**LAG**) of the difference.

Hempel's project, narrowly construed, sets the priorities this way: First provide a necessary condition for scientific explanations of facts and then try to provide criteria or at least a necessary condition for contingent generalizations to be laws. It might look like a kind of circularity is involved if we thought that the concept of a law in turn required an appeal to explanation. Nevertheless, we propose an account, a mini-theory (**LAG**), of two concepts: laws and accidental generalizations. The theory provides necessary conditions for both concepts, each of which involves an appeal to explanation. Accordingly the mini-theory, and its consequences seem surprising, given the simplicity of the proposal.

Basically the idea in an early form is, I think, due to the German Nineteenth Century philosopher Hermann Lotze, who distinguished between true universal conditionals, and accidental ones. I have called that idea the Lotze Uniformity Condition (**LUC**). Our theory has two parts:

(1) If a contingent generalization is a law, then there is some explanation of all its instances, and
(2) If a contingent generalization is accidental, then there is no one explanation R, of all its instances.

In Chap. 7, we will provide simple arguments for some of the consequences of this mini-theory:

(i) No law is an accidental generalization.

[2]Recall that physical magnitudes are functions, and functions (say those of one argument for example) are relations R(x, y) such that if R(x, y) and R(x, z), then x = z.

(ii) All laws are true. i.e. the factivity of "It is a law that ..." This assumes that explanation is factive in the sense that "A is a part of the explanation that B", (E[A; B]) implies A as well as B.
(iii) The statement that S is a law (L(S)) is non-extensional., assuming that explanation is non-extensional.
(iv) If S is any contingent generalization that is explained, then S is not accidental.

This last conclusion is particularly nice for it provides an answer to the question of why we should want to explain some generalizations. Isn't it enough that the generalization is true? (iv) provides a compelling answer. It's good to know that S is well-confirmed, and even better to know that S is true. But an explanation tells us even more about the world: its truth is not accidental.

Finally, we consider what this theory makes of three examples usually cited in the literature, as non-laws. There is H. Reichenbach's "All gold cubes are less than one cubic mile." (G), E. Nagel's "All the screws in Smith's car are rusty." (N), and Peter Hempel's "All the rocks in this box contain iron."(H). Towards that end we propose a refinement of the Lotze uniformity condition, so that instead of requiring that there is a uniform explanation of the instances of any generalization that is a law, we require that also the explanation be a theoretical one – i.e. there is some theory that explains all the instances of the law.

We therefore have (**LUCT**), the Lotze Theoretical Uniformity Condition, rather than (**LUC**). A similar adjustment is made for the condition on accidental generalization. It turns out that in these three cases, the verdict is the usual one: none of them are laws.

In laying out the claims of both (**LUCT**) and (**LAGT**), we do not by any means think that they are all that can or should be told about laws and accidental generalizations. We think they are just part of a fuller story. Also, although it is true that on the basis of this account so far, there is no explaining an accidental generalization, it must be stressed that this result does not diminish their scientific importance, utility, or even their use in explaining other generalizations.

Having now introduced the idea of explanation by theories in our discussion of the mini theory (**LAGT**), we will now focus on some issues involved in the idea that sometimes, theories explain laws.

**Part II**: Schematic and non-schematic theories, two kinds of explanation, the physical modals provided by theories, and the nomic modals associated with laws.

**Chapter 8** concerns the accounts of Ernest. Nagel and Richard B. Braithwaite of the explanation of laws by theories. These views unfortunately have not been recognized for their radical departure – each in its own way- from the so-called "standard view". Nagel's view, in our reconstruction of it, developed the idea that some theories that explain may be *schematic.* That is, they don't have a truth-value. Braithwaite on the other hand, explicitly restricted his discussion of laws and their explanations to those that are embedded in what he calls "well-evidenced deductive systems". Thus the key notions for Braithwaite are "is a law of the well evidenced deductive system D", and "is an explanation of the law L in a well-evidenced deductive system D∗". The effect is that the concepts of a scientific law, and the explanation of laws by theories are relativised to associated deductive systems. Both these kinds of innovative departures have had an impact on the present study that will become evident in our consideration of the neglected cases of important significant schematic theories and our focus on those laws that are relativized to theoretical scenarios.

Although Nagel and Braithwaite owed much to Norman Cambpell's account of the structure of scientific theories and the explanation of laws, their accounts are radically different from each other, and from anything that preceded their proposals.

We hasten to add that the decisive influences on their theories are very different. In Nagel's work, the influence of the work of D. Hilbert and M. Pasch's work on Geometry is very evident. We argue that in fact the outcome for Nagel's account of laws and their explanation leads to the conclusion that some theories that explain are *schematic*, and consequently, neither true nor false. Needless to say that that requires some adjustments to the view of how explanations can take place – given that schematic theories seem to violate the factivity condition for explanations. Even Nagel's claim that all explanations are deductive (even probabilistic ones), has to be reexamined. So, we think that Nagel's account is radical, and of great interest.

In the light of Hilbert's influence on Nagel's and our own account, we will devote the following chapter to Hilbert's stunning view that *all* properly axiomatized theories are schematic. This, Hilbert thought, was true for all theories, mathematical and physical. His reasons were clearly explained in his correspondence with G. Frege. That explanation was mangled however, by an unfortunate translation into English. It is clear that Hilbert meant that all theories, mathematical and physical are schematic, and as a consequence, neither true nor false. The details of this fascinating proposal are therefore the topic of Chap. **9**, Hilbert's Architectural Structuralism, and schematic theories.

Braithwaite's account has an interesting source as well, and marked an important radical divergence from the Hempelian-like account that preceded it. It was F. Ramsey who initially considered a aproposal (due to J.S. Mill) which he formulated compactly this way:

> Laws are consequences of those propositions which we should take as axioms if we knew everything and organized it as simply as possible in a deductive system.

Ramsey rejected it for two reasons.

> That it is impossible to know everything and also to organize it in a deductive system.

David Lewis is credited with proposing a widely popular emendation of the Millean view, in which he expunged the epistemic feature:

> ... a contingent generalization is a law of nature if and only if it appears as a theorem (or axiom) in each of the true deductive systems that achieve a best combination of simplicity and strength. [3]

[3] D. Lewis, Counterfactuals, Harvard University Press, Cambridge, Massachusetts, 1973, p. 73.

Braithwaite offered, in our opinion, an even better response. He required that his laws, and the explanation of laws be embedded in what he called well-evidenced deductive systems – with the possibility that the laws and their explanations would be embedded in different well-evidenced deductive systems, This relativization to well-evidenced deductive systems was innovative, and, as we shall show, it enabled Braithwaite's account to yield a number of very surprising deductive conclusions. Unfortunately, he never made them explicit.

**Chapter 9**, Hilbert's Architectural Structuralism and Schematic Theories. A discussion of D. Hilbert's view about theories may come as a complete surprise, since Hilbert is known as a world-class mathematician with contributions of the first order (no pun intended) in mathematics and logic. One would expect a sophisticated series of penetrating remarks about mathematical or logical theories; not physical ones. Nevertheless, Hilbert held views that are relevant to matters of physical laws and their explanations. His thoughts were forcefully and clearly articulated by him in a correspondence with G. Frege. Unfortunately the correspondence was badly translated in a widely circulated English translation of the correspondence that mistakenly translated the key word "Fachwerk" by "scaffolding".

Hilbert vigorously defended the idea that every properly axiomatized theory, mathematical or physical, was a Fachwerk of concepts, but the translation had it as "Every theory is a scaffolding of concepts." Obviously a scaffolding that is used in a construction is discarded once the construction is completed. Why would anyone want to discard the theory? Up until the 1930s at least, Hilbert and his students used the term "Fachwerk" frequently, when they wrote up the lectures for their seminars. The term was used to describe a type of architectural construction (timber-frame constructions come close to what is meant). One difference is that timber-frame constructions are usually covered over and remain hidden. Not so with Fachwerk, where the construction is not to be concealed, but viewable. The concepts of a theory he says, are the "infilling" of the interstices of the Fachwerk, and the infilling can occur in an infinite number of ways, subject to the constraints set by the Fachwerk.

We propose that the best way to think of theories as Fachwerk is to think of them as schematic theories. This ties in nicely with the remarks that Hilbert often made to the effect that theories are neither true nor false. Hilbert even provided a proof that there are infinitely many ways of filling the schema.

Schematic theories however, raise very interesting questions. How can such theories explain anything, since they are neither true nor false? Moreover, although schematic theories have many applications, it is clear that some of the applications that are subsumed under them are laws. Since we will show that those items that are subsumed under schematic theories are not deduced from them, then an interesting question arises. If the laws that are subsumed under schematic theories are not deductive consequences of those theories, then one has to understand that these successes of those theories are not among its deductive consequences. The restriction of the successes of theories to include only their deductive explanatory and predictive consequences therefore needs to be qualified.

The next two Chaps. **10, and 11** are devoted to developing an account of modality in those cases when there is a well-defined collection and the members of that collection are its possibilities.

**Chapters 10** and **11** are focused on a new account of the *physical* modal operators that theories provide, and a subset of those modals that are associated with laws – the *nomic* modal operators. Chapter **10** begins with a survey of a variety of scientific laws, each associated with a background theory, or theoretical scenario, in which some physical magnitudes figure. Sometimes the laws are consequences of that scenario, but not necessarily so. It is the theoretical backing that will be used to show that theories provide modal operators, as well as modal entities.

The laws are usually expressed with the aid of magnitudes that are drawn from the associated theoretical scenario. The focus of this essay is on those laws that have associated theoretical scenarios. Some examples: Newton's three laws of motion, his law of gravitation, Kepler's laws of planetary motion, Galileo's law of falling bodies and Ohm's law. In addition, we also will survey a host of physical theories that use widely different examples of states to indicate the possibilities that those theories provide for the behavior of systems that those theories describe.

It is fairly common to think that those states, in all their variety, are all described as possibilities. This is true not only for simple cases, such as the elements of a probabilistic sample space, where the outcomes of experimental trials, are described as possibilities. The same talk of possibilities is often used in the discussion of powerful physical theories that in which reference to the states of a system in a certain theory tell us about what is possible, and what is not possible for those systems, given certain conditions. The theoretical scenarios are so varied, so that it is clear that there is not one concept of state shared by all. Nevertheless those states are all thought of as possibilities.

We think of these theories or theoretical scenarios as providing a concept of a state of that theory. The rest of Chap. **10** is devoted to explaining the notion of a magnitude vector space that is associated with each theory, and it is the basic elements of these magnitude spaces that provides for each theory, the set of physical possibilities provided by the theory. It will happen that there are different vector spaces for different theories. However each of these vector spaces are constructed in the same way out of the particular physical magnitudes of the theory – only the magnitudes may be different. So the first part of our account is that each theory provides a set of physical possibilities, and those possibilities are just the basic states of the magnitude vector space that is associated with the theory.

The serious question is of course this: simply callings these states possibilities may just be a case of speaking with the vulgar. We want to show that the appeal to the states of a theory is an appeal to modal operators and modal entities that that are genuinely modal. In the following chapter we will show that there are many important examples, physical theories included, where items are described colloquially as possibilities, and the description is not just a loose way of speaking, but an appeal to a new kind of modality.

**Chapter 11** opens with a discussion of various examples of uses of possibilities in the sciences: probability sample spaces where the outcomes of an experiment (a toss of a die, say) are the possibilities, cases in talk of truth-values as possibilities, modal talk of possible worlds, physical possibilities like the two paths between two points in a configuration space, possible cases in a proof, and possibilities provided by a physical theory.

We next introduce the idea of an *implication structure* (any non-empty set together with an implication relation on it). In each of these cases one can define an associated implication structure and the possibilities are of two kinds: modal operators, defined as operators **on** that structure, and modal entities that are special members **in** that structure.

We begin with the simple example of a probability sample space and then describe an associated implication structure for it. On that structure we provide explicit definitions of a necessity operator (box) and a possibility operator (diamond) as special functions that map the set S of the structure to itself. It is shown that these operators are very close to being like S5 modal operators. In addition to these operators that are functions ***on*** the implication structure, we also show how to define those possibilities (modal entities) that are elements ***in*** the set S of the implication structure We obtain the same result if we consider a physical theory T, rather than a probability same space. As in Chap. **10** we construct the magnitude vector space (MVT) associated with it. And given this space, we then form the implication structure IT, associated with it (using a construction similar to the one used to associate with a probability sample space an associated implication structure). In the way we treated the probability case, so too in the cases of any physical theory, there are necessity and possibility modal operators close to C.I. Lewis's system S5, that are operators ***on*** the implication relation IT, and there are modal entities in that implication relation. Finally, we have already defined a special class of modal operators, *Gentzen structural modal operators*, and we prove that all the physical modal operators provided by a physical theory are *Gentzen structural modal operators*. So not only are they close to S5, but they also belong to a class of modal operators to which almost all the familiar modal operators belong. That is simply another way of saying that they are not exotic, but belong to a broad class of modal operators and are thereby located in a modal landscape.

**Chapter 12.** In previous chapters we noted that for some laws, their theoretical scenario is a schematic theory. We propose that good examples of such schematic theories are ready to hand: Newton's theory of mechanics, Lagrange's theory of mechanics, Hamiltonian theory, and Kolmogorov's theory of probability. We maintain that they are schematic. Granted that we are correct in this, there are some deep changes that have to be made to familiar doctrine: (1) It is no longer true that all theories have a truth-value because, because some theories are schematic. This is different from the one-time instrumentalist claim that no theories have truth values because they are rules. (2) Given (1), it is no longer true that theories can explain laws. The reason is that explanations are factive, (we assume that they are) and schematic theories are neither true nor false. (3) It is not possible for theories to be explained by other theories, again for failure of factivity. Thus any claim of reduction of one theory to another is impossible, if reduction of one theory by another is an explanation. (4) The familiar claim that all explanations of laws by theories are deductive, has to be qualified. We maintain that that there are many cases, including major conservation laws, and even Newton's Law of inertia that are subsumed under schematic theories, where the relation of subsumtion is different from deducibility. The result is that these laws count as major successes of those schematic theories, but they are not among the deductive consequences of those theories. To cover such cases, we introduced (in Chap. 9) the notion of *subsumptive explanation* (explains∗) for schematic theories according to which schematic theories can explain∗ laws, and "explains∗" is factive. Finally we indicate some ways in which explains and explains∗ are systematically related, and tentatively investigate, (with the aid of a mini-theory (**GTL**), the requirement that if a theory explains a law, then the theory is at least as general than that law.

# Chapter 2
# Hempel's Deductive-Nomological Model: In the Beginning . . .

I shall assume for the present that the basic features of Hempel's Deductive Nomological Model of Explanation[1] are familiar to the reader[2]: a deductive explanation of some fact Ga about a particular object a, requires some other fact Fa, and since it was assumed that Fa would not by itself yield a deduction of Ga, that the deductive connection between the two facts was to be supplied by an additional premise L –a law. So, though the target of their paper was scientific explanation, we want to consider what it tells us about scientific laws.

They had arrayed a number of further conditions which we will neglect here. We will try to show that although they thought that laws were universally quantified material conditionals, the model, the *Deductive Nomological Model* that they provided yielded a more general conclusion: scientific laws *imply* the universally quantified material conditional. The idea was that any deductive explanation of a fact by others would have the form

$$\frac{\begin{array}{c}\mathrm{Fa}\\ \mathrm{L}\end{array}}{\mathrm{GA}}$$

Though the point is sometimes contested, it seems clear that the Deductive Nomological Model was intended only as a necessary condition on scientific explanations.

The law that is involved in the explanation is supposed to enable a deduction from one fact to another with its aid. Laws, on this account have an inferential role.

There is, we suggest, one underlying assumption which clarifies the situation somewhat. This assumption, which I shall call the *Lotze Uniformity Condition* (to be studied in more detail in Chap. 7), requires that the law (whatever that may be) should provide for an explanation not only for why a is G, given, that it is F; that law

[1] Carl G. Hempel and Paul Oppenheim, "Studies in the Logic of Explanation," *Philosophy of Science,* Vol. 15 (1948).

[2] C.G. Hempel and Paul Oppenheim, "Studies in the Logic of Explanation", *Philosophy of Science,* Vol. 15, pp. 135–178.





A. Koslow, *Laws and Explanations: Theories and Modal Possibilities*, Synthese Library 410, https://doi.org/10.1007/978-3-030-18846-7_2

should also be part of the explanation of why any other b is a G, given that it is also F. That is, the law should be part of a uniform explanation for all of its instances. This condition, or something close to it, is the basis we think for a distinction between laws and accidental generalizations, and we shall return at a later point to a fuller discussion of the distinction between statements that are accidental rather than non-accidental.[3] It is a substantial requirement that requires some argument, as we shall see, since it involves a departure from a long-standing tradition of how laws and explanations are related.

The "tradition" has it that accounts of scientific explanation ought to rely on some notion of scientific law. We shall propose instead that in order to distinguish laws from other generalities, there has to be a reliance on some concept of explanation.

We have been a little circumspect about the kind of logic that Hempel required for the deductions to take place. Nevertheless, we shall assume that however sophisticated or plain the logic might be, it includes two claims. The first is that for any A, B, and C,

(1) If A, B together imply C, then A implies (If B, then C), and
(2) If U implies Fa, and "a" has no occurrences in U, then U implies the universal generalization ∀xFx.

The first condition is known as the Deduction Theorem, and it is sometimes assumed as part (but only a part) of the definition of the conditional (A → B). The second claim is usually known as the rule of inference UG (Universal Generalization). With these two assumptions in place, we can advance the argument a bit further:

Thus far we have that L, and Fa together imply Ga. By (1) we then have that L implies the conditional (Fa → Ga). Consequently, by (2), we have the result that the law L implies ∀x(Fx → Gx), provided of course "a" doesn't occur in L.

Thus whatever other features the law L might have, it has to imply the universal generalization. Well, L has to do at least that much, and it looks as if Hempel chose the weakest of all the possible candidates for being a law – the universal generalization of a material conditional.

The universal generalization certainly provides a logical connection between the antecedents and consequents of all its instances. That is, it provides a reason, for all conditionals (Fa → Ga), in fact *the same reason*, for all those instances – though it may not be the same explanation of all the instances. There is a difference, which we shall look into when we discuss the Lotze uniformity condition in more detail (Chap. 5).

Thus far we have a motivation for explanations of a rather limited sort: explanations of one fact such as Ga, by another, Fa. Within this narrow compass, the representation of laws by universally quantified conditionals has a certain cogency. However it's only for a very special kind of case, and even within these limits, there are serious problems with it –as the literature amply testifies. Many of the objections to this model stem from a difference over whether the model is supposed to be a

[3]Cf. Chap. 7

necessary condition for deductive explanations, a sufficient condition, or both. We shall not be concerned with that kind of criticism.

What we have in mind are two critiques, one by F. Dretske (and independently M. Tooley), and the other by D. Armstrong,[4] that cut to the heart of the Deductive Nomological Model. They initiated a new surge of interest in developing a philosophical account of scientific laws, and an incentive to develop non-Humean accounts of them. Both attacked Hempel's deductive model for its crucial assumption that scientific laws were best represented by universally quantified material conditionals.

Dretske claimed that laws could not be represented as universally quantified material conditionals because the substitution of any coextensional predicate for any predicate in the conditional would preserve the truth of the conditional, but some of those substitutions would result in a conditional shifting from a law to a statement that was not a law. The conclusion was that although universally quantified material conditionals are extensional, laws are not. So laws are not universal conditionals. We shall consider that argument in the next chapter.

Dretske also considered the claim which N. Goodman (and R. Chisholm) championed: Laws support their corresponding counterfactuals. However, in the present situation, clearly universally quantified conditionals do not support their corresponding counterfactuals –again leading to the conclusion that on their own, these quantified conditionals do not express laws. And to make the difference between laws and universal conditionals even more dramatic, Dretske maintained that laws were not even general; they were singular statements. All of this will be considered and rebutted in Chap. 4.

Armstrong however attacked the Hempelian representation of laws in another way: laws record necessitations: It is not just a matter of fact that metals expand when they are heated; laws have a kind of necessity to them. The idea is that laws are physically necessary –that is, there is a special kind of modality which holds of them. That, as we shall note, is a familiar idea that goes back probably to Aristotle, and more recently, to William Kneale and others. Armstrong sometimes described the law that all metals expand when heated, in an emphatic way as "All metals *must* expand when heated". This is to be contrasted with the flat claim that all metals (in fact) expand when heated, The modality (if it is one) seems to reside for Armstrong in the way in which the universal metal brings forth, or (as he says), *generates* or brings about the universal (if it is one) of expansion (upon heating). Armstrong as we shall see, argued that the best representation of laws is not as universal conditionals with predicates in the antecedent and consequent positions. Instead, a law is a binary[5] relation (which he called "necessitation") between

[4]F. Dretske, "Laws of Nature", *Philosophy of Science*, 44, 1977; M. Tooley, "The Nature of Laws", Canadian Journal of Philosophy, 7, 1977; D. Armstrong, "What is a Law of Nature?", Cambridge University Press, Cambridge, 1983.

[5]The emphasis of one universal generating another seems to be behind Armstrong's requirement that laws are binary relations. The fact that fairly simple laws such as the gas law that pressure times

universals. There are deep-seated reasons why this account of laws fails, that have to do both with a faulty assumption about predication, and the inadequacy of universals to express scientific laws. These considerations we shall postpone for Chap. 6.

Before we turn to Dretske's account of laws, it is worth noting that the original Hempelian model –at least in the form we have provided for it, is not without resources in the light of these criticisms of it.

We have described the Hempelian view of laws inferentially, as the missing deductive link between deductive explanation of one fact Ga by another Fa. The law was supposed to deductively connect not only those two particular facts, but other similar facts such as Gb and Fb, and G(c), and F(c). That is, a certain uniformity was assumed to be provided by laws. The analogy of this kind of uniform deductive connection that was supposed to be provided by a law might have been thought as expressing logically, the intent of invoking a causal connection, only more clearly, and more generally by logical means. Though I hasten to add that I am not aware of any place where Hempel was thinking of the logical connection as somehow reflecting a causal connection.

The general motivation for Hempel's model obviously leaves it open that there are many candidates for this inferential and explanatory role, and we saw that Hempel chose the weakest of all of them. However, that isn't the only possibility. Instead of the universally generalized material conditional $(\forall x)(Fx \rightarrow Gx)$, there is the possibility of using the counterfactual $(\forall x)(Gx \;\Box\!\!\rightarrow Fx)$ instead, – a universally quantified counterfactual conditional. Let us suppose then that laws can be represented as counterfactual conditionals, and suppose further that we adopt as a theory of counterfactuals, the system (**VC**), David Lewis's favorite logic of counterfactuals.[6]

Thus we are about to explore the possibility of representing laws as counterfactual conditionals using Lewis' (**VC**) account of them. We hasten to add that Lewis does not endorse a counterfactual representation of laws. His account – a *systems account*, eschews any use of modal or other non-extensional idioms.[7] Lewis did not endorse a counterfactual account of laws. That however does not prevent anyone interested in buttressing Hempel's Deductive Model by going counterfactual and appropriating Lewis's favorite theory of counterfactuals for that purpose.

The resultant advantages of modifying Hempel and Oppenheim's model by going counterfactual are evident. First, it does provide a deductive inferential link between facts and the universal quantification of a material conditional. The reason is that in

volume equals a constant times the temperature (PV = nRT) involves three magnitudes, raises some technical issues that surely have to be addressed.

[6]D. Lewis, *Counterfactuals,* Harvard University Press, 1973, p. 132.

[7]"a contingent generalization is a *law of nature* if and only if it appears as a theorem (or axiom) in each of the true deductive systems that achieves a best combination of simplicity and strength)", *Counterfactuals,* p. 132. We will return to this kind of systems account in Chap. 8.

Lewis's system (**VC**), the counterfactual A□ → B implies A → B. And so we think we will also have (∀x)(Fx □ → Gx) implies (∀x)(Fx → Gx).[8]

In fact the use of quantified counterfactual conditionals, as we shall see later, would block the complaints of both Dretske and Armstrong that the standard classical quantified material conditional was too weak because laws were not always preserved under the substitution of coextensional terms. The truth of counterfactuals however, isn't always preserved under such coextensional substitutions.

The claims of Hempel, Goodman and a host of others that one way of distinguishing between those generalizations that are laws and those that are not is that only laws imply (or support) their corresponding counterfactuals is now trivially true. Of course laws support their corresponding counterfactuals; they *are* counterfactual conditionals. You cannot have a simpler solution.

A second type of objection to the laws of Hempel's deductive model was that there was a necessity that laws have, but no trace of it is to be found in universally quantified material conditionals. However, as we shall see, if laws are to be represented by quantified counterfactual conditionals, then it will become evident that such statements (the counterfactual conditionals) are themselves statements of necessity. That would meet the demands of those who saw laws as modals of some kind, and it would be a serious objection against those who thought that there is no modal character possessed by laws. So the move to a counterfactual representation of laws would decide issues of necessity or no necessity for laws, but it would not settle the issue as to what type of necessity was furnished by counterfactuals – i.e. was it physical, metaphysical, or neither.

A third kind of objection could also be met by the move to the counterfactual: Hempel often emphasized the claim that explanations, deductive or probabilistic, show that what was explained was to be expected. It might be argued that representing a law – say that all metals expand when heated as the counterfactual: "If anything were a metal, then it would expand when heated", shows better than the simple quantified conditional, that the heating of the metal was to be expected. This point is a slender one. It rests on the suggestion that if we knew that anything which was a metal would expand on heating, then we would expect that that if it were to be heated then it *would* expand. It's a slim advantage at most, over the straightforward universal conditional "If anything is a metal then it *will* expand upon heating". The counterfactual conditional states that it would expand (not that it will). I think that the "will" reading has a slight edge, but the difference is not enough to favor a counterfactual representation of scientific laws.[9]

---

[8]One further caveat: (**VC**) does not cover quantified sentences. We shall assume that for the quantificational cases above, the natural generalization will work. Therefore the quantified counterfactuals can piggy back on whatever the universal conditionals can do deductively. However, the formulation of a system for quantified counterfactuals has some technical difficulties, since it turns out to be a kind of quantified modal logic, and there is at present no one standard account of quantified modal logic.

[9]Since the quantified counterfactual about the expansion of the metal implies the indicative quantified material conditional in Lewis' system (**VC**), it would seem to follow that the quantified

Thus far we have been examining the viability of using counterfactual conditonals as a way of expressing scientific laws, and rescuing the Hempelian treatment of them in his account of scientific explanation. I think however that there are considerations that show decisively that on balance, the efforts to save the Hempel model by going counterfactual are ill-conceived.

There are unfortunate consequences of going counterfactual for the representation for scientific laws - both for the theory of counterfactuals and for scientific practice.

Consider for example Newton's first law, the law of inertia. According to it, if there are no forces acting on a body, then, if at rest, it will stay at rest, and if in motion with velocity v, then it will continue to move in a straight with that velocity. More briefly,

(1) If there are no forces acting on a body, then it will not accelerate.

That's the usual way in which the law is stated.

The philosophical revision under consideration however, suggests a counterfactual formulation. So, (1) will be represented as the counterfactual (2)

(2) If there were no forces acting on a body B, then that body would not be accelerating. i.e.
$\neg F \,\Box\!\!\rightarrow \neg\, ACEL(B)$.

Now it is commonplace to also express the law of inertia in an equivalent, but contrapositive way, – in fact the way that J.C. Maxwell [*Matter and Motion*} and much later R. Feynmam [*The Feynman Lectures*] expressed the law, we have:

(3) If a body B is accelerating, then there is a force acting on it.

The counterfactual representation of (3) is given by (4):

(4) $ACEL(B) \,\Box\!\!\rightarrow F$,

If a body B were accelerating, then there would be a force acting on it.

The problem now is this: The formulation of the Law of Inertia (1), and its contrapositive (3) are regarded by physicists as logically equivalent formulations of the law of inertia. If we express this equivalent form of the law counterfactually, then we should have that (2) and (4) are logically equivalent. Consider the one-way implication of (3) to (1).

What are the consequences if we assumed that for any A and B that

(5) $A \,\Box\!\!\rightarrow B$ implies $\neg B \,\Box\!\!\rightarrow \neg A$,

Then it follows that for any A, B, and any C, that

(6) $A \,\Box\!\!\rightarrow B$, implies $(A \wedge C) \,\Box\!\!\rightarrow B$.

counterfactual yields not only the expectation that the metal *would* expand upon heating, but that it also yields the expectation that the metal *will* expand upon heating. That strange result might lead to the belief that it is not a good idea to read off expectations from conditionals, counterfactual or indicative.

The proof is straightforward: Suppose that $A \Box\!\!\rightarrow B$. By (5) we have $\neg B \Box\!\!\rightarrow \neg A$. Now considering the system (**VC**) of counterfactual logic, we conclude that for every C, we have that $\neg B \Box\!\!\rightarrow (\neg A \vee \neg C)$ ((**VC**) has a rule of closure: if E counterfactually implies F, and F logically implies G, then F counterfactually implies G). By (5) we then have $\neg(\neg A \vee \neg C) \Box\!\!\rightarrow \neg\neg B$. Since negation is assumed here to be classical, and for counterfactuals substitution of logically equivalent sentences for logically equivalent sentences results in logically equivalent sentences, we conclude that (6) for any A, B, and C. That is bad news. It is a well-known fallacy of counterfactual reasoning. For example, suppose that if Albert were to go to the party, that he would have a wonderful time. That doesn't imply that if Albert were to go the party in a coma, he would have a wonderful time.

I chose the Newton's Law of Inertia because it and its contrapositive are used as equivalent expressions of that law. I think that this is so for any law, expressed as a conditional. It and its contrapositive are ways of expressing the same law. Even in the mundane case of "All metals expand when heated" the same seems to hold. The counterfactual version, "If anything were a metal, then it would expand upon heating" is equivalent to the contrapositive "If anything failed to expand upon heating, it wouldn't be a metal."[10]

The general conclusion we think is just that in the case of laws that are expressed as conditionals, they and their contrapositives are equivalent, and the counterfactual versions are not. Our argument, to recapitulate it, is that in any case when a counterfactual ($A \Box\!\!\rightarrow B$) is equivalent to (or even implies) its contrapositive, then that counterfactual implies $(A \wedge C) \Box\!\!\rightarrow B$ for any C. And for each conditional law, that is not correct. We think that what is true of the Law of Inertia is true of all conditional laws. It is correct to reason from the conditional to its contrapositive. However if conditional laws are represented as counterfactuals, it is not correct to reason logically from the counterfactual representation of the law to the counterfactual version of its contrapositive. If you insist that (conditional) laws be represented as counterfactuals, then the practice of physics says that you have assumed that that two counterfactuals are logically equivalent when they aren't. Or you can draw the conclusion that since it is not correct to reason from a counterfactual to its contrapositive, then it is not correct to argue that conditional laws are not equivalent to their contrapositive. In the first case you run up against what is believed to be a counterfactual fallacy, and in the other case you run up against well entrenched reasoning in physics. Best to drop the idea that laws should be expressed as counterfactuals.

Of course this is a claim only for those laws expressed as conditionals.[11] In the non-conditional cases, the notion of contraposition makes no sense.

We have seen that the counterfactual representation of laws (of the conditional sort) is deficient as representations go. But that is not the end of the matter. There is another way in which counterfactuals have been used, other than as representations

[10]This is not so if the logic used is intuitionistic.

[11]If a law is not expressed as a conditional, then of course all talk of its contrapositive makes no sense.

of laws. They have figured as one way of differentiating between those generaliztions which are laws from those which are not. This is obviously a different role for them than that of representation. And it is the subject of the next chapter.

# Chapter 3
# Laws and Corresponding Counterfactuals, – An Untenable Connection

In the preceding chapter, we explored the possibility of how serious criticism of the Hempel model of explanation might have been met by assuming that laws were representable as counterfactuals. That assumption only made matters worse.

We now wish to look at a weaker claim that has had many advocates: It is the idea that laws are distinguished from true contingent generalizations in that they support their corresponding counterfactual conditionals, while mere contingent generalizations do not. We think that even this weaker connection is untenable.

There are two widely endorsed criteria that have been used to distinguish those generalizations that are laws from the rest. One is that laws support their corresponding counterfactual conditionals, – *the counterfactual connection*, and the other is that laws possess a special kind of necessity that non-laws don't have. These two criteria are closely connected, but neither one is intended to be both a necessary and sufficient condition. In this chapter we will consider only the counterfactual connection, and we will postpone modal considerations for a later chapter.

The counterfactual connection for laws is simply that laws support their corresponding counterfactual conditionals. The notion of "*support*" needs some explanation. It cannot mean evidential support. There can be evidence for laws, but there are almost no cases where some laws are evidence for, or confirm other laws. I suggest, following the few leads in the literature, according to which "support" of the corresponding counterfactual means "logical implication". Thus we understand the counterfactual connection to be the claim that

(C) Laws imply their corresponding counterfactuals.

We will refer to (C) as *The Counterfactual Connection*. The immediate problem is that this formulation is ambiguous. There are at least two possible readings:

(1) If A is a law, then A implies its corresponding counterfactual conditional. and
(2) $\mathcal{L}$(A) implies the counterfactual corresponding to A,
where "$\mathcal{L}$(A)" just states that it is a law that A.


A. Koslow, *Laws and Explanations: Theories and Modal Possibilities*, Synthese Library 410, https://doi.org/10.1007/978-3-030-18846-7_3

These can be recast a bit more formally, to render the ambiguity more apparent. If A is a conditional, then let CC (A) be the counterfactual conditional corresponding to A. In the case that A is a material conditional (If B then C), then its corresponding conditional is (If B were true, then C would be true), i.e. $B\Box\rightarrow C$. (1) and (2), then are respectively

(1∗) If $\mathcal{L}$(A), then A $\Rightarrow$ CC(A).
(2∗) $\mathcal{L}$(A) $\Rightarrow$ CC(A).

The first says that if A is a law, then it is A that implies the corresponding counterfactual, while the second requires that the corresponding counterfactual follows from the statement "A is a law". There is clearly a difference between "A" and "It is a law that A". If "It is a law that A" implies A, then (1∗) implies (2∗). However we do not assume that this is so.

(2∗) is interesting for two reasons. When combined with the claim, which we shall defend below, that if A is a conditional, then its corresponding counterfactual conditional implies A, that is

(3∗) CC(A) $\Rightarrow$ A,

we then obtain the result that

$\mathcal{L}$(A) $\Rightarrow$ A – that is,

If any conditional is a law, then it is true, assuming (2∗) and (3∗). Moreover, given (3∗) it follows that (1∗) implies (2∗). So there are these differences between (1∗) and (2∗) that are worth exploring.

Either (1∗) or (2∗) can be used to deny that something is a law on the grounds that it doesn't support its corresponding counterfactual. Consider Hempel's well-known example "All the rocks in the box have nickel in them". It is not a law, he claimed, because the corresponding counterfactual is false.[1] If (2∗) is used, then, we conclude that it's false that A is a law. And if (1∗) is used, then the implication A $\Rightarrow$ CC(A) is false since A is true and the corresponding counterfactual is false, and again, by (2∗) it is false that A is a law. A $\Rightarrow$ CC(A).

We can see these two readings are embedded in the literature. For example Psillos[2] in his astute discussion of the connection between laws and counterfactuals states the assumed connection this way:

> It is a law that all Fs are Gs if and only if (i) all Fs are Gs, and (ii) if an object x had been an F it would also have been a G. (146).

This implicates (1∗). On the other hand, Hempel seems to have had (2∗) in mind when he wrote:

> The sentences "All members of the Greenbury School Board for 1964 are bald.", and "All pears in this basket are sweet." illustrate this point. Goodman has pointed out a characteristic

[1]Carl Hempel, *Philosophy of Natural Science,* Prentice-Hall, 1966, p. 56.
[2]S. Psillos, *Causation and Explanation,* McGill- Queens University, Press 2002.

> that distinguishes laws from such nonlaws: the former can, whereas the latter cannot, sustain counterfactual and subjunctive conditional statements. Thus the law about the expansion of gases can serve to support statements such as 'If the oxygen in this cylinder had been heated (were heated) under constant pressure then it would have expanded (would expand)'; whereas the statement about the School B lends no support at all to the subjunctive conditional 'If Robert Crocker were a member of the Greenbury School Board for 1964 then he would be bald'. (339)

It seems fairly clear that Hempel thought that if A is a law then it, A, implies the corresponding counterfactual.

Nevertheless, despite these two different ways of understanding the so-called counterfactual connection, the distinction does not get picked up in the uses to which the connection has been put. One example may suffice to indicate how easily the distinction can go undetected.

Suppose our concern is whether Nagel's example (S) "All the screws in Smith's car are rusty" is a law.[3] If it is a law, then by (1∗) it follows that "All the screws in Smith's car are rusty"(S) implies that if anything were a screw in Smith's car then it would be rusty. That implication fails on the grounds that S is true but the corresponding counterfactual is false. Consequently, it is not a law that all the screws in Smith's car are rusty. That same conclusion follows if we used (2∗) rather than (1∗). For the counterfactual about the screws in Smith's car is false, and so by (2∗), it is not a law that all the screws in Smith's car are rusty. Therefore either version of the counterfactual connection, (1∗) or (2∗), will deliver the result that "All the screws in Smith's car are rusty." is not a law.

Despite the fact that the two versions of the counterfactual connection yield the same verdict, there are different dramatic consequences for the adoption of one or the other. One might adopt (1∗), as many writers do, with the idea that it provides a modest way of distinguishing laws from other generalizations. We say "modest" because initially, it doesn't require, as does the representational view, the equivalence of a law with a counterfactual. All that is required is that the law itself *implies* the corresponding counterfactual.

Nevertheless there is a dramatic consequence of (1∗), that we believe to be incorrect. We will argue that from (1∗) it follows that not only does the law imply its corresponding counterfactual; it is equivalent to it. Thus every law is equivalent to a counterfactual conditional.

To see how the proof of this goes, we have to explain a bit more about corresponding counterfactuals. Suppose that A is either a material or a counterfactual conditional. The notion of a corresponding counterfactual arises when the contrast is drawn between a material conditional and its counterfactual corresponding to it. Consider, for example "If Plato has no beard then it was Ockham who shaved it" ($P \rightarrow O$). Given any A, we will indicate the corresponding counterfactual to A by CC(A). Thus $CC(P \rightarrow O)$ is $P \Box\!\rightarrow O$. That is, "If it were the case that Plato has no beard, then it would be the case that It was Ockham who shaved it". We can summarize this idea as follows:

[3] E. Nagel, *The Structure of Science*, Harcourt, Brace, & World, Inc., 1961, p. 52.

Confining ourselves then to two kinds of conditional, material and counterfactual, we shall say that

(i) If A is the material conditional (B → C), then CC(A) is (B□ → C,)
(ii) If A is the counterfactual conditional (B□ → C), then CC(A) is (B□ → C), and
(iii) If A is neither a material nor a counterfactual conditional then CC(A) is A.

Thus, material conditionals get converted to the obvious counterfactual, counterfactuals are left intact, and statements that are neither, are left intact. (3*) follows easily then that for all A, that is,

$$CC(A) \Rightarrow A.$$

The reason is that if A is a material conditional (say B → C), then by (i), and the fact that in most systems of counterfactual logic, including D. Lewis' (**VC**), we have $B\Box \rightarrow C \Rightarrow (B \rightarrow C)$. If A is a counterfactual then by (ii),CC(A) is just A. If A is neither a material nor a counterfactual than by (iii), CC(A) is again just A.

This result however, is bad news if we are trying to choose between the two readings of the counterfactual connections, (1*), or (2*). For, if we assume (1*), then, If $\mathcal{L}$(A), then A ⇒ CC(A). Since CC(A) implies A, we then have as a consequence that If $\mathcal{L}$(A), then A ⇔ CC(A). That is, all laws are equivalent to counterfactuals.

Thus, although we have assumed something that is apparently weaker than the representational thesis that (conditional) laws are counterfactuals, we wind up with the same unwelcome conclusion that we had when we assumed the representational thesis, that all laws are counterfactual.

There are some advantages, and some disadvantages to this conclusion. -One decisive disadvantage as we noted in the previous chapter is that this conclusion. Runs squarely up against scientific practice. On the other hand, if you think there is something modal about laws, then the claim that laws are counterfactuals provides some support.[4]

The representation of laws as counterfactuals has a nice feature, – if you think that there is something modal about laws. It has been argued that laws are not only true generalizations; they have some kind of necessity. If it is granted that there is something modal about counterfactuals, then although we haven't established that there is some necessity to laws; we have shown that there is some modal character that they do have.

It should be noted in passing, however, that there are some writers who have thought that laws being counterfactuals is not an unwanted consequence but a

[4]It was argued by Victor H. Dudman, "Interpretations of 'If' – sentences." in *Conditionals*, F. Jackson (ed.), Oxford Readings in Philosophy, Oxford University Press, 1991, pp. 202–232, that counterfactual conditionals are such that with the exception of antecedents that are theorems or contradictions, the counterfactual is equivalent to a modal operator associated with the antecedent that is applied to the consequent.

desirable one. For example, W.E. Johnson in his *Logic*[5]made it a practice to cast all the laws under discussion in his three-volumes of Logic, as counterfactual conditionals.

On balance, we think, the consequence that the representation of laws as counterfactuals runs up against scientific practice is a decisive mark against it.

The prospects for the second version of the counterfactual connection, (2∗) look much brighter. We have already noted that in connection with the plausible condition (3∗) above, it follows that all (conditional) laws are true. This has usually been claimed to be a feature of all laws with very weak arguments in its support. Hempel thought it would be strange to say that laws weren't true.

There is a stronger case to be made for the truth of laws by relying on the assumption that all explanations are factive, so that any law that is involved in any explanation would have to be true. Of course this does not cover the case for all laws, unless you assume that every law is involved in some explanation. That is one huge assumption, and I have never seen the case made for it.

The second way of understanding the counterfactual connection, (2∗) yields some very nice consequences about laws. (2∗) yields the conclusion that all (conditional) laws are true. We noted that (3∗) was true. So (2∗) together with (3∗) yields $\mathcal{L}(A) \Rightarrow A$.

Another very interesting consequence of (2∗), concerns non-extensionality of "It is a law that". It provides a simple proof of the following:

"It is a law that . . ." is a non-extensional context.

That is, the substitution of coextensional predicates in the sentence "A" will not always preserve the truth value of "It is a law that A". Assume (2∗), that $\mathcal{L}(A) \Rightarrow CC(A)$, and that it is also the case that the corresponding counterfactual is non-extensional. The argument then goes as follows: Suppose that A is a law. So it's true that $\mathcal{L}(A)$. Therefore by (2∗), CC(A) is true. Now CC(A) is non-extensional. So the result of substitution of *some* coextensional predicates for those in A will yield a sentence A∗, such that CC(A∗) is false. But, by (2∗), we have $\mathcal{L}(A*) \Rightarrow CC(A*)$. Consequently $\mathcal{L}(A*)$ is false. That is, A∗ is not a law.

Of course, it would be very interesting if "$\mathcal{L}$" were not only non-extensional, but also modal. That, is, as far as I can tell, still an open question, which we shall try to settle positively in Chap. 11.

It should be clear by now that "$\mathcal{L}(A)$" and "A" are not equivalent. If they were then by (2∗), it follows that "$\mathcal{L}(A)$", "A", and "CC(A)" would all be equivalent to each other. Consequently all laws would then be equivalent to counterfactuals, – and that we already argued in Chap. 2, is not so.

We have argued that the traditional attempts to single out laws from other generalizations by using their corresponding counterfactuals fails. That does not

[5]*Cambridge University Press*, Part I, 1921, Part II, 1922, Part III in1924, Reprinted by Dover Publications, Inc., 1964.

entirely settle the matter whether the use of some counterfactuals other than the corresponding ones cannot be put to use to sort out the laws from other generalizations.

One recent attempt of interest has been provided by Marc Lange.[6] It is a rather intricate and nuanced attempt that uses counterfactuals that are different from those that we have called corresponding counterfactuals, so that our previous arguments against the usual uses of corresponding counterfactuals do not apply. Nevertheless, there are other problems with Lange's account of the relation between laws and counterfactuals. Lange endorses the fundamental principle (NP), according to which,

> m is a law if and only if in any context, $p\Box\!\!\rightarrow m$ holds for any p that is logically consistent with all of the n's (taken together) where it is a law that n (that is to say, for any p that is logically consistent with the first-order laws. (20).

Lange's account organizes laws into a hierarchy which we will not go into at present., with the first-order laws at the bottom. Let LL be the set of all the first-order laws. We assume in the following that the set LL is consistent. Otherwise there wouldn't be any p that is logically consistent with LL, and In that case NP would be vacuously true and not of much interest.

Now assume that m and m∗ are two laws of LL. Since m∗ is consistent with LL, it follows that $m*\Box \rightarrow m$. Also, since m is consistent with LL, we have that $m*\Box \rightarrow m$.

Consequently, any two first order laws, m and m∗ are counterfactually equivalent. That is, $m*\Box \rightarrow m$, and $m\Box \rightarrow m*$, for any two first-order laws. That doesn't seem to be correct, either for any two laws of a theory (say the First and Second laws of Newtonian Mechanics), or laws from two different theories (say Genetics and Quantum-Electrodynamics).

There is a second consequence of (NP) which is also somewhat troubling. Let p and q be two first order laws and so members of LL. Each of them is consistent with L (again assuming that LL is consistent). Form their disjunction $p \vee q$. It too has to be consistent with LL. For suppose otherwise. Then the negation of the disjunction would follow from LL. It follows then that the negation of p and the negation of q each follow from LL. In that case then, LL is inconsistent since it implies p as well as its negation (the same for q and its negation). The result then is that (NP) implies that the disjunction of any two laws is a law. Again, a highly dubious conclusion.

There are various ways of blocking these conclusions. One could drop the requirement that LL, the set of all first-order laws is consistent. That is not a plausible adjustment. One could appeal to some notion of relevance which Lange does do in various places, by requiring that the context will provide some idea of what counts as an appropriate antecedent (like the p used in the expression of (NP), and also supply some argument to show why the use of first-order laws in LL are not appropriate antecedents in the counterfactual, and why the use of disjunctions of laws are not

[6]*Laws and Lawmakers, Science, Metaphysics, and the Laws of Nature,* Oxford University Press, New York, 2009.

appropriate antecedents of the counterfactuals appealed to by (NP). In any case, some further refinement of (NP) is needed.

We have explored an unnoticed ambiguity in a familiar way of connecting laws with their corresponding counterfactuals so that laws could be distinguished from other contingent generalizations. On one reading (1∗) it becomes evident that there is a conflict between what it says, and what scientific practice teaches. On the other reading, (2∗), it becomes evident that the prefix " It is a law that..." implies that the statement that it is a law that (A) is factive ("It is a law that A" implies A.), and also that "I is a law that A" is nonextensional: for some substitution of coextensional predicates that occur in A, the truth-value of "It is a law that A" will change in truth-value. This is an important difference between "A", and "It is a law that A", and its neglect results in a good deal of mischief in the account of laws of Fred Dretske which we will discuss in the next chapter.

# Chapter 4
# F. Dretske's Total Rejection of the Hempel-Model

## 4.1 The Challenge

Dretske's influential account of laws stands in sharp contrast to the classical accounts of Hempel, Nagel, Carnap, and Braithwaite. I will, in the following, refer to these somewhat different accounts as "the standard view".[1] Dretske's view has been seen as marking a striking shift away from that view, and the beginning of a series of new non-Humean accounts.[2] His account of laws forcefully declared that

(1) Laws are relations between, properties, quantities, magnitudes, or features that are expressed by certain predicates.
(2) The relations in each law are extensional, and can vary from law to law,
(3) Laws are singular sentences rather than general, and
(4) The statement of the law All F's are G will not change it's truth-value if any co-referential expression replaces "F", "G", or both. However the result of that replacement may not be a law. This feature is, he, misleadingly refers to as "the opacity of laws".

With the exception of the first claim, the remaining ones appear to be a wholesale repudiation of the received view about laws. They are not friendly amendments. If true, they cut deeply into the received view. We turn therefore to a detailed consideration of them.

[1]The "standard view" is clearly a convenient misnomer. There are several views that it covers, that are radically different from each other, but more akin to each other than to those critics and critiques that followed the view of Hempel.

[2]F. Dretske, "Laws of Nature", *Philosophy of Science* 44, 2, June 1977, pp. 248–268.


A. Koslow, *Laws and Explanations: Theories and Modal Possibilities*, Synthese Library 410, https://doi.org/10.1007/978-3-030-18846-7_4

## 4.2 Laws as Relations

It is worth quoting a seminal passage (i) in its entirety to get a sense of the detailed argument. We shall divide that passage into three contiguous parts, so that what looks like a very radical departure from the standard view can be more easily delineated. Consider the statement (A):

(A) All Fs are G.

Apparently Dretske thought that (A) could not be understood as a universal truth and also as a law, only a different one. It is not at all clear to me that (A) is ambiguous. Nevertheless the clear implication is that no law is a true universal conditional. The difference is explained this way:

> (i) To conceive of (A) All Fs are G as a universal truth is to conceive of it as expressing a relationship between the extensions of its terms; to conceive of it as a law is to conceive of it as expressing a relationship between the properties (magnitudes, quantities, features) which these predicates express (and to which we may refer with the corresponding abstract singular term).

Dretske says that, as a universal truth, A uses a relation and, as a law, A is also uses a relation, but the relations in each case are different. However it's just incorrect to think that both universal material conditionals, eg. $(\forall x)$ $(Fx \subset Gx)$,[3] and laws involve relations. Universal conditionals involve the material conditional which is a connective; not a relation. We shall see below that this difference can be rectified by a small adjustment.

Moreover, the predicates "Fx" and "Gx" are not extensions of predicates, but the predicates themselves. Dretske's description of this case seems to differ from the customary usage. Nevertheless, again, there is a small repair that one can make, the result of which is that there is not a serious difference between true universal conditionals and laws over the use of different relations.

One thing Dretske might have had in mind is the extensions of the predicates "Fx" and "Gx", – i.e. $F*$ and $G*$ which are the respective sets (extensions) of those predicates. In that case the universal truth might be expressed as $(\forall x)$ $(F* \subseteq G*)$, where "$\subseteq$" is indeed a relation between $F*$ and $G*$- the *set-theoretical* relation of inclusion. This is indeed a universal statement which describe a *relation* between the two sets $F*$ and $G*$. The two statements $(\forall x)$ $(Fx \subset Gx)$ and $(\forall x)$ $(F* \subseteq G*)$ don't have the same meaning, though they mutually imply each other, given a certain amount of elementary set theory, and some assumptions about when and if certain predicates have extensions. So it is not all that clear that this particular distinction that Dretske wants to draw between universal conditional truths and laws is all that

[3]The horseshoe indicates the material conditional and the single shafted arrow is used by Dretske in the representation of laws, which may be used for a variety of different relations used in the representation of various laws. (253)

sharp. Even the Hempelian account can represent laws as relations between extensions with a little help from set theory.

There is a second set of differences between laws and universal material conditionals. These differences drive home the deep difference Dretske wants to make between laws and universal conditionals. Each law is supposed to be a relation that holds, not between predicates, but between certain properties, which he seems to refer to indifferently as the magnitudes, quantities, or features, which those predicates express. The expressions of laws, he maintains, use abstract singular terms (F-ness and G-ness) to refer to the properties related by a law.

In the second part, of the paragraph, (ii) below, Dretske proposed yet another way of marking the distinction between laws and universal conditionals. Laws, he claims, do not provide an extensional context for the terms that are related by the law. As he succinctly states it: "laws involve extensional relations between intensions." (263) The extensionality of the various relations in laws is different from an important feature of laws that Dretske thinks is an important feature of laws: they are opaque. We turn next to his explanation of these new differences.

The terms F-ness and G-ness are the universals that correspond to the predicates "Fx" and "Gx". He provides an argument involving diamonds and kimberlite to support this important difference between universal truths, and laws. We shall for the moment postpone discussion of that argument in order to explain the type of relations that laws involve. In (ii) below, Dretske argued that that every law has the form:

(L) F-ness ➨ G-ness,

where the relation "→" is supposed to be an *extensional* relation (sic) between the properties F- ness and G-ness. Dretske says that the single shafted arrow is to be read as "yields" of "generates".[4]

By the extensionality of the relation (in L), he means that "any term referring to the same quality or quantity as F-ness, can be substituted for it in (L) without affecting it's truth or it's law-likeness" (263)

This point about the extensionality of a relation is a bit obscure. In the sentence (L) there is a relation between two terms F-ness, and G-ness, each of which is thought to refer to a property, universal, or physical magnitude, and, were we to substitute some other term for F-ness that is coreferential with it, then the truth-value of (L) would remain unchanged.[5]

Dretske claims that the relations of various laws are all extensional, but that laws are opaque. The term 'extensional' is used by Dretske in a non-standard way.

---

[4]If "yields"or "generates" is meant, then there are two serious problems. (1) Those terms seem to be binary relations. What if more than two magnitudes or universals are expressed in the law?, and (2) The notion of yielding or generation sees to require that the law relates magnitudes at different times so that laws that relate magnitudes at simultaneous time are excluded.

[5]I assume here that Dretske has something specific in mind in speaking of coextensive properties, universals, or magnitudes. It's not obvious what the coextensiveness of two properties or universals, or physical magnitudes comes to.

Opaqueness of laws we think is usually a way of expressing the idea that the statement "It is a law that A" is not extensional. That is, the truth-value of "It is a law that A" is not preserved under substitution of terms for others that are coextensive. Here is his argument that for claim that laws are opaque.

> (ii) The opacity of laws is merely a manifestation of this change in reference. If "F" and "K" are coextensive, we cannot substitute the one for the other in the law "All F's are G" and expect to preserve truth; for the law asserts a connection between F-ness and G-ness and there is no guarantee that a similar connection exists between the properties K-ness and G-ness just because all F's are K and *vice versa.*

Some care has to be taken in assessing Dretske's claim for the opacity of laws. This becomes a bit easier if we consider his particular argument about diamonds mined in kimberlite., in support of his important claim for that opacity.[6]

> Consider a predicate expression "K" (eternally) coextensive with "F"; i.e., (x) (Fx ≡ Kx) for all time. We may then infer that if (x) (Fx ⊃ Gx) is a universal truth, so is (x) (Kx ⊃ Gx). The class of universal truths is closed under the operation of coextensive predicate substitution. Such is *not* the case with laws. If it is a law that all F's are G, and we substitute the term "K" for the term "F" in this law, the result is not necessarily a law. If diamonds have a refractive index of 2.419 (law) and "is a diamond" is coextensive with "is mined in kimberlite (a dark basic rock)" we cannot infer that *it is a law* that things mined in kimberlite have a refractive index of 2.419. Whether this is a law or not depends on whether the coextensiveness of "is a diamond" and "is mined in kimberlite" is *itself* law-like. The class of laws is not closed under the same operation as is the class of universal truths. (250)

We think that Dretske's argument for the opacity of laws falls short.

> Suppose that "Diamonds have a refractive index of 2.419" is a law. If we replace "is a diamond" by the coextensive "is mined in Kimberlite" the result is not a law. The result of this substitution however, does not show that "Diamonds have a refractive index of 2.419" is a non-extensional idiom. What the example shows is that "*It is a law that* Diamonds have a refractive index of 2.419" is a non-extensional (opaque context). What his argument illustrates more generally, is that the statement "A is a law", is an opaque context for the predicates in the sentence A, but A itself may not be an opaque context for the predicates that occur in it.

This is an important insight, but it not news to the advocates of the so-called standard view. We have already noted, in Chap. 3, that the non extensionality of $\mathcal{L}$(A) is an easy consequence of the claim (2∗) that "It is a law that A" implies the corresponding counterfactual. We argued there that the counterfactual connection in (1∗) is incorrect. However, as we shall argue below, in Chap. 7 ("Laws and Accidental Generalizations"), the non-extensionality of "$\mathcal{L}$(A)" is an easy consequence of a necessary condition that we will provide in that chapter for "It is a law that . . .".

---

[6]There is bigger game at stake than just a radical view of laws. If laws are non-extensional, which their opacity entails (the usual idea is that a context provides an opaque context if and only if it provides a non-extensional context) then one would have to abandon the Quinean program of eschewing non-extensional idioms from scientific discourse. Science, replete with laws, on Dretske's view would be a huge source of serious non-extensionality – the laws. In what follows, we will use non-extensional context" and "opaque context" interchangeably.

This would certainly not be news for advocates of the standard view. Most advocates of a Hempelian kind of account used something like (2∗) to distinguish between laws and accidental generalizations (Chap. 7). The opacity claim doesn't bring anything new to what earlier accounts were already committed. Most of the earlier accounts claimed that there was a close deductive relation of laws to corresponding counterfactuals. And if that connection is deductive, then the statement that some statement A is a law ($\mathcal{L}$(A)) would be non-extensional. But it does not follow that A is non-extensional.

In short, the law "Bodies not acted upon by any force move uniformly." is not opaque. However, "*It is a law that* bodies not acted upon by any force move uniformly." is opaque.

Briefly then, "$\mathcal{L}$(A)" is opaque (non-extensional) shouldn't be confused with the claim that "A" is non-extensional. Thus Dretske's idiosyncratic account of the opacity of laws is not a radical repudiation of the familiar view.

In the passage (iii) below, Dretske emphasized two points that appear to be in striking contrast to familiar assumptions about laws.

> (iii) "It is this view that I mean to defend in the remainder of this essay. Law-like statements are singular statements of fact describing a relationship between properties or magnitudes. Laws are the relationships that are asserted to exist by true law-like statements. According to this view, then, there is an *intrinsic* difference between laws and universal truths. Laws imply universal truths, but universal truths do not imply laws. Laws are (expressed) by *singular* statements describing the relationships that exist between qualities and quantities; they are not universal statements about the particular objects and situations that exemplify these qualities and quantities." (253–4).

We now want to consider in greater detail Dretske's claims that (1) laws are singular statements, and (2) that they are relations involving universals or intensional terms, and (3) that the relations between those intensional terms can vary from law to law. These three claims are not minor adjustments to proposals commonly found in standard accounts of laws; they are a wholesale rejection of them. We consider these claims in turn.

### 4.2.1 *Laws as Singular Statements*

There are two salient features of the law-like relations which Dretske has emphasised in this seminal paper. The first is that although the relation involved in various laws may be different from law to law, the relation in each case is extensional. The second feature requires that whatever the relation may be, it relates "intensional entities". These entities Dretske usually refers to as universals. He doesn't indicate exactly what he means by that notion, but it is clear that physical magnitudes are included as a special case.

The first difference between the received view and Dretske's is the seemingly innocent one: that laws are relational statements, which, as already noted is not the way the received view represented them, according to which they are universally quantified material conditionals. In contrast the target of Dretske's discussion are sentences of the form

F-ness ➡ G-ness.

We can nevertheless minimize this difference between the account of Dretske and the received view acccording to which laws are just true universal material conditionals of the form

(A) : $(\forall x)(Fx \subset Gx)$.

With only some minor technical adjustments, we can bridge the difference between the two versions: Instead of using the predicates "Fx" and "Gx" in (A), let $F*$ be the set of all things x, for which "Fx" holds, and similarly let $G*$ be the set of all things which satisfy "Gx". Thus, $F*$ and $G*$ are sets – the extensions of the predicates, and let $\subseteq$ be the relation of set inclusion. Then (A) can be expressed as

$(A^*)(F^* \subseteq G^*)$.

That is, everything which is in the set $F*$ is in the set $G*$. We now have a version of (A) that expresses a relation (set theoretical inclusion) between two specific items, two particular sets.

"($F* \subseteq G*$)" is a singular statement that relates two particular sets. Dretske seems to think that the fact that on his account laws are singular statements signifies an important departure from the received view that laws are general. However the distinction between singular and general in this case is exaggerated. ($A*$) is a singular statement, but it is also equivalent to a general one. It states that one set is a subset of another. Consequently the fact that laws according to the older view, are not relations, and are general, is not in my opinion a significant difference between Dretske's account and the older one. With simple modifications, using elementary set theory, the standard account can easily bridge that difference.

This reply will not completely meet Dretske's challenge. It does not meet Dretske's claim that laws are relations between intensional terms. We now will address this more refined view of Dretske.

## 4.3 Laws as Singular Statements that Relate Intensional Terms

Suppose, as Dretske maintains, laws are singular statements that relate universals, or as he sometimes says, they relate intensional entities (sometimes he appeals to the notion of physical magnitudes instead). Consider an example of a law that he considers at length:

(1) All diamonds have a refractive index of 2.419.

Assume, for the sake of the argument, that this is indeed a law about diamonds. Suppose further that some background minimal theory U of universals is assumed.

We shall presently say more about how minimal U needs to be. We also assume along with Dretske that there are universals associated with the predicates "x is a diamond" ("Dx") and "has a reflective index of 2.419" ("Rx"), "being a diamond", and "being something that has a refractive index of 2.419"respectively, or "D-ness" and "R-ness" – or just "D" and "R" for short. Then let Inst(D) and Inst (R) be the set of all instances of D and R respectively. We assume that if there is going to be some appeal to universals, then the concept of the instances of them is also part of that minimal theory, and since universals are regarded as intensional entities by Dretske, so too are the collections of instances of universals. For, the entrance condition of say Inst(D) is the condition of being an instance of the universal D-ness. We assume as part of our minimal theory, that if U is any universal, then the set of all its instances exists (compatible of course with the universal having no instances). We have to be careful about this blanket assumption. It can, without proper qualification lead easily to a Russellian paradox, – if we allow for a universal whose instances are all those universals that are not instances of themselves (that is, X is an instance of R if and only if it is not the case that X is an instance of X). So we shall assume that suitable restrictions are incorporated in the minimal theory to block the Russell-like paradoxes. Such qualifications would be needed for any theory of universals, so it's a reasonable desideratum for a minimal theory.

Assuming all this, then the law that all diamonds have a refractive index of 2.419 tells us at least this: every instance of D is an instance of R. That is,

(2) $\text{Inst}(D) \subseteq \text{Inst}(R)$.

Now define an extensional relation S for any universals U and V, as follows:

(3) $S(U, V)$ if and only if $\text{Inst}(U) \subseteq \text{Inst}(V)$.

"S(U,V)" is extensional in this sense. If for any universals U∗ and V∗ such that U and U∗ have the same instances, and V and V∗ have the same instances, then S(U,V) if and only if S(U∗,V∗).

This relation S between the universals D and R is an extensional relation between the sets Inst(D) and Inst(R), which are intensional entities.

What this seems to show is that given this background of a minimal theory of universals, every law of the type All F's are Gs is an extensional relation between particular intensional entities. Therefore Dretske's claim that his view is contrary to the standard view, is not a problem for a resourceful Humean who adopts some minimal theory of universals. With the addition of a minimal theory of universals, the result is not that much different from what the received view had already endorsed.

In fact we can press matters a bit further. Suppose that it is also assumed that for every *predicate* "Fx" used in expressing a law, there is a corresponding universal $F^*$ such that

(4) For any x, Fx if and only if x is an instance of $F^*$ (x is in the set $\text{Inst}(F^*)$) that is, $\forall x(Fx \leftrightarrow x \, \varepsilon \, \text{Inst}(F^*))$.

That is, we assume that the xs that fall under a predicate *used in a law* are just those that are instances of a corresponding universal.

Of course it is not true on most theories of universals that to *every* predicate, there is a corresponding universal. However, (4) does not assume such a general correspondence. It requires a correspondence only for those predicates that occur in some expression of a law. Such an assumption already seems tacit in Dretske's view that the correct understanding of the law that All F's are G's, does not involve the predicates "Fx" and "Gx", but their associated universals F-ness (F*) and G-ness (G*). The law is best presented, according to Dretske, not as a universal conditional involving the predicates "Fx" and "Gx"; it is best represented as relation between their associated universals F* and G*.

Assumption (4) does not imply that two universals will be identical if and only if their corresponding sets of instances are identical. It is compatible with the possibility of two distinct universals having exactly the same objects fall under them.[7] In fact one might even take (4) to be a way of expressing the condition under which a predicate has an associated universal. However, in Dretske's diamond example, there seems to be no question that there are universals that are associated with each of the predicates "is a diamond" and "has a refractive index of 2419". So, including (4) as part of the minimal theory, we can make the following simple point:

The position that laws are extensional relations between intensional entities doesn't pose an insuperable challenge to the older standard Humean view that laws are universal conditionals between suitable predicates. The Humean account can express laws like All Fs are G as an extensional relation between intensional entities. In particular, the law about diamonds would be expressed as S(D, R), where S is the inclusion relation between the sets of the instances of the universal D and the universal R. That is, Inst (D) $\subseteq$ Inst (R). However, this is equivalent, assuming (4), to the universal conditional $\forall x(x \in \text{Inst}(D) \rightarrow x \in \text{Inst}(R))$. Now assuming that (4) holds for the predicates "Dx", and "Rx" and their associated universals D and R, it follows that

(5) Inst (D) $\subset$ Inst(R) is equivalent to $\forall x(Dx \rightarrow Rx)$.

The result then is this: someone who adopted the minimal theory of universals that we have outlined, and a principle like (4) which just says that those objects that satisfy the predicates of laws are exactly those objects that fall under a corresponding universal, could then argue that laws like "All diamonds have a refractive index of 2.419." could also be described by a singular extensional relation between intensional entities. And in fact, that singular statement would be equivalent to the universal conditional

(6) $\forall x(Dx \rightarrow Rx)$.

Which is of course just the way that an unrepentant Humean would have it in the first place.

[7]To that extent then, it is different from Fregean concepts, which are identical if they have exactly the same instances.

Thus given an elementary bit of a theory of universals, it would follow that Dretske's proposal that laws are singular statements involving extensional relations between intensional entities is not an advance over the received view of laws. It's not far off from that older received view– only dressed up with the help of a very minimal theory of universals. In the end, the older view is equivalent (given an elementary theory of universals) to what the Humean maintained all along.

### *4.3.1 Are There Many Relations That Are Used in Various Laws?*

According to Dretske's representation of laws, the law that all Fs are Gs is represented by "F ➡ G", (F-ness generates or yields G-ness), where the relation "(X generates Y)" is indicated by the single-shafted arrow, and the universals of the law are indicated by "F-ness" and "G-ness". It is clear that Dretske believes that although the single-shafted arrow represents the generation relation of laws, it need not stand for the same relation in every law. For him, the relation of one universal or magnitude yielding or generating another could be different for different laws. Athough the relations might be different, they presumably would still be relations of the generative or yielding type –whatever that means.

There is however, another type of law that Dretske discusses, other than the conditional kind like his diamonds example (1977, p. 253, fn. 7). It involves equations that relate magnitudes to each other. His example is the nice one of Ohm's Law (1977, p. 253, fn. 7). Ohm's law is usually expressed in the form of an equation: $E = RI$. E is the impressed electromotive force (voltage) in a closed circuit, R is the resistance of the circuit, and I is the current. The more accurate version takes into account that once the circuit is closed and the current increases there is an electromotive force opposed to the impressed voltage given by LdI/dt where L is a constant of the circuit, – the coefficient of self-induction. Consequently, the correct equation at any time t is given by $IR = E - LdI/dt$, so that $I = E/R (1 - Rt/L)$[8] at any time t,

Here we have a law which Dretske expresses in the conventional way as an equation between several physical magnitudes. Since many serious scientific laws are expressed as equations, it would seem obvious that those laws too are also expressed as relations between magnitudes.

For the present we will think of magnitudes as being at the very least functions that map physical entities (particles, fields, etc.) to the real numbers, or to the elements of some suitable mathematical structure –vectors, tensors, spinors, etc). We know that if $\mu$ and $\rho$ are magnitudes, that not every mathematical function of them will also be a magnitude. For example, the product of mass and position is not a

[8]Cf. R. Courant, *Differential and Integral Calculus,* Interscience Publishers, 1949 v.I, 182–3, 433–5.

magnitude, and although time is a physical magnitude, it is clear that the exponentiation of it, $e^t$, is not. Nevertheless, *some* mathematical functions of magnitudes will also be magnitudes, and these are usually called *functionals* – because they are functions of functions. In this way we can see that the more accurate form of Ohm's law, $I = E/R\,(1-e^{-Rt/L})$, is also a relation between magnitudes. The left hand side of the equation is the magnitude I, and on the right hand side, we have the magnitude (functional) $E/R\,(1-e^{-Rt/L})$.

It is fair to say that there are a host scientific laws that are represented as equations between physical magnitudes. However the relation between the magnitudes is just the identity that holds between functions. Since functions are normally taken in extension –that is functions are identical when their values are identical for each argument, it follows that the relation between magnitudes in these equational type of laws is just the identity relation that we have between numbers, vectors, tensors, or spinors. Since the functions are taken in extension, it's essentially the identity of sets that is needed.

Therefore, contrary to Dretske's claim, the law-like relation between the universals or magnitudes doesn't vary from law to law. It's always the same: it's the identity relation between functions. That economy is a plus. Moreover, the relation as we have described it is an extensional one. That too supports Dretske's account. And it also supports his claim that the law is singular in the sense that is says of two specific magnitudes that they are identical.

The crucial place where the present account differs from Dretske's is that what are related in laws are not intensional entities; they are magnitudes or functions of a certain kind, like speed, mass, length, and electronic charge. They are the usual familiar functions with nothing intensional about them.[9]

### 4.3.2 *What About Non-equational Laws?*

We have seen how one of Dretske's examples, Ohm's Law, leads to an account that shows that the relation of that law is just the identity relation between functions. The relation in all laws that are of this equational sort will also involve the identity relation. Consequently, among this large class of laws there's no variation of relations from law to law of the kind that Dretske says there is.

What then about the kind of law that is conditional and not equational, such as

(1) All diamonds have a refractive index of 2.419.

[9]The identity relation between magnitudes is extensional, but requires some care. The mass m of an object a can be either a number, r, or a multiple of some unit. I.e. m(a) = r grams, or m(in grams (a) = r. In either case, identity between magnitudes will be extensional: the same objects get the same values.

Suppose, along with Dretske, that this is a law. Then according to him it should relate physical magnitudes. Now we have assumed that physical magnitudes are mappings, usually from objects to real numbers (or to vectors, tensors, or other kinds of mathematically tractable entities).[10] If so, then although it is admittedly a bit of a stretch, it is easy enough to think of "diamonds" and "having a refractive index of 2.419" as magnitudes. One device that can be used is the mathematical notion of a characteristic function. So for "diamond" associate the function $\varphi_D$ which assigns to any object the value 1 if that object is a diamond, and has the value 0, otherwise. Similarly associate to "has a refractive index of 2.419" the function $\varphi_R$ which assigns 1 to any object that has a refractive index of 2.419, and has the value 0 otherwise. Thus

$\varphi D(x) = 1$ if x is a diamond, and 0 otherwise
$\varphi R(x) = 1$ if x has a refractive index of 2.419, and 0 otherwise.

Each of these characteristic functions is a magnitude – to be sure at the low end of magnitudes since they have only the two values of 0 and 1. We might call them *meager magnitudes,* or *discrete magnitudes*. With these elementary definitions in place it follows that (1) is equivalent to

(2) For every x, [ if $\varphi D(x) = 1$, then $\varphi R(x) = 1$].

However it is easily seen that (2) is equivalent to

(3) For every x, [$\varphi D(x) = \varphi D(x) \| \varphi R(x)$][11]

With (3) we have an expression of the law about diamonds and refractive indices now expressed as a law in which magnitudes are related by an identity relation between them.

It is now apparent that the examples of laws that Dretske used to argue for a radical departure from the received view of laws, all seem to be adequately represented as relations between magnitudes, with the same identity relation used in all cases.

We note in passing that by this use of characteristic functions, it is also true that every universally quantified material conditional can be shown to be equivalent to an equation between characteristic functions. The argument is just the one that we have given. Therefore even "All ravens are black" can be expressed as an equation, using characteristic functions for the predicates "is a raven", and "is black".

It might be objected, with some justice, that the characteristic function $\varphi_B(x)$ for the predicate "is black" (i.e. for which $\varphi_B(x) = 1$, and 0 otherwise) if and only if x is black), is not a magnitude, not even a meager or discrete one. The characteristic function is very close to what Frege took to be a concept. Concepts on Frege's account of them, are functions that map objects to the truth values t, f. They are close

[10]This assumption about magnitudes will be studied and developed in Chap. 10.

[11]In (3) there is the product of two magnitudes. This we have called a functional and here we assume that this functional is also a magnitude.

cousins of characteristic functions that are mappings from objects to 1, 0. In fact in the computations just given, we could have used functions which map to the truth values t, f, rather than 1, 0. It would become reasonable then to think of such conditional laws as relations between Fregean concepts.

The use of Fregean concepts, however, would beg the question against Dretske, since they are extensional. Unlike other kinds of concepts, Fregean ones are identical if and only exactly the same objects fall under them. By using "magnitudes", which is the terminology Dretske himself used, there's no begging of the question against him.

In summary, we have seen that a close analysis of Dretske's use of two laws,

"All diamonds have a refractive index of 2.419" and Ohms Law does not support his claims that laws are singular statements (and not generalizations), that they are opaque (not truth preserving under substitutions of terms with coreferential ones), and that they employ a variety of generating relations (one universal generating another) to relate universals (rather than using a uniform extensional notion of a mathematical notion of function relating two (or more) magnitudes.

We will devote the following two chapters to understanding what went wrong, and what went right with DavidArmstrong's account of scientific laws. Chapter 5 (*Prelude to Armstrong. A mathematical debate, and the philosophical implication for the expression of laws (B. Russell, F. Ramsey, and D. Armstrong*)).

It describes a revolution in mathematical physics in the Nineteenth century, that centered around the concept of function in mathematics and physics. It involves a shift from an intensional concept of function to an extensional concept. There is good circumstantial evidence to think that the debate over the transition was known to Frank Ramsey, and was the basis of one of his criticisms of Russell and Whitehead's *Principia*. Essentially it was that the *Principia* used intensional propositional functions, and unless they shifted to an extensional notion, their claim to have reduced elementary arithmetic to logic would fall short of that task. More precisely, using only an intensional notion of function, their efforts would be descriptively incomplete – e.g. that a set had only two members could not even be defined.

What we wish to show in Chap. 6 (*Identity lost, gained, and lost again*) is that Armstrong's account of scientific laws as a neccessitation relations between universals is subject to the same kind of criticism that Ramsey made against Russell and Whitehead.

# Chapter 5
# Prelude to Armstrong: A Mathematical Revolution That Inspired F. Ramsey, and Left Russell and Armstrong Unmoved

The story here involves F. Ramsey's realization that the nineteenth century mathematical debate about functions had implications for the expression of statements of arithmetic in Russell and Whitehead's *Principia*. We believe that it is the same flaw, – expressive inadequacy – that lies at the heart of what is wrong with D. Armstrong's account of scientific laws.

The debate concerned a shift to the now familiar concept of functions as arbitrary assignments from sets to sets (the *extensional* concept of a function), rather than adherence to the more restricted concept of function, according to which the values assigned had to conform to a rule or condition that restricted the arbitrariness of what was assigned (call it the *intensional* concept of a function). F. Ramsey, we believe, was aware of the discussion that had earlier raged on the continent, and saw that contemporary philosophers (Russell and perhaps Wittgenstein) were using a notion of propositional function that was intensional rather than extensional. He called Russell out over the use of intensional propositional functions in the first edition of the *Principia Mathematica* which, he claimed was a major defect of that project. The reason was that the exclusive use of such functions did not yield a satisfactory identity relation. Consequently Russell would not be able to retrieve Peano Arithmetic given the weak resources of the theory he relied upon. We believe that D. Armstrong's account of laws is subject to a very similar defect, which will be explained in the following Chap. 6.

The non-extensional concepts of propositional functions, and relations have great interest in various philosophic fields. Nevertheless, the restriction to using only non-extensional concepts of functions and relations exclusively in science and mathematics results in seriously limiting both the mathematics and the science that can be expressed and explained. The cost is enormous for those disciplines, and we think it is also enormous for the philosophies of those disciplines.[1]

[1]It's the *exclusive* used of non-extensional functions that gives rise to severe problems. Many sciences and even classical logic are not so resrtricted. For example classical probability is a clear





A. Koslow, *Laws and Explanations; Theories and Modal Possibilities*, Synthese Library 410, https://doi.org/10.1007/978-3-030-18846-7_5

The transition from the older to our contemporary account of functions involved a remarkable change which was fought over in two arenas: mathematical physics from the mid-to-late nineteenth century, and mathematical logic of the early twentieth century. In the former, the conflict concerned the use of functions taken extensionally; in the latter it concerned the use of relations and propositional functions in extension (as they were called by F. Ramsey).

In mathematical physics one of the issues at stake was the adequacy of the theory of non-extensional functions for the solution of problems of mathematical physics. F. Ramsey's challenge was to the adequacy of Russell and Whitehead's account of current mathematical theories such as arithmetic and analysis on what they took to be logical grounds. In both cases, scientific and philosophical, the outcome favored the extensional accounts of functions and relations. Consequently, we view the return to the older nonextensional notion of function, propositional and mathematical, as a return to a much earlier time, and scientifically a step backward. This may not be a drawback for those who wish to use non-extensional propositional functions and relations in metaphysics, or the philosophy of language, or indeed any other field of philosophy where these notions are relevant and of significance.

For example, B. Russell developed a theory of direct reference with the use of propositional functions (mappings of objects a, to propositions P(a)), where any singular proposition, P(a) is an ordered pair, <a, P>, of an object a and a property P). This notion of propositional function is at bottom a use of a non-extensional concept of propositional function. We certainly do not wish to discount the importance of such a notion of proposition. What we want to emphasize is that huge amounts of contemporary science and mathematics could not be expressed were we restricted to the use of this particular notion of proposition and the notion of propositional function that is based on it. The effect on a philosophy of science that wants to square it's results with scientific and mathematical practice would be enormous, were its notion of scientific law restricted by this adherence to non-extensional resources.

We shall try to show (in the following chapter), that at root, some current accounts of laws, in particular that of D. Armstrong harken back to older, quite powerful intuitions that favored the adherence to non-extensional relations and functions.

Before proceeding, it is worth explaining what we take to be the differences between extensional and non-extensional propositional functions and relations.

---

case because the probability function is non-extensional. In the case of classical first-order logic, it has been argued that the quantifiers are non-extension al. Cf. A Koslow, "The modality and non-extensionality of the quantifiers" *Synthese*, 2014.

## 5.1 Extensional and Intensional Relations

**(a1) Extensional Version** The usual extensional notion of an n-ary relation R on a set S, is that it is a subset of the Cartesian product $S^n$: S x S x. . .x S of S with itself (n times). That is, the set of all ordered n-tuples $<x_1, \ldots x_n>$, with all the members of the n-tuple belonging to S.

The extensional account of relations requires two principles:

(1) *Every* n-place relation on S is a subset of $S^n$, and
(2) *Every* subset of $S^n$ is an n-place relation on S.

The non-extensional account of relations can, on most versions of the notion, accept (1). It is with the rejection of the second condition that the nonextensionalist parts company with the extensional view. On the extensionalist view, it follows that any two n-place relations on S will be subsets of $S^n$, and so will be identical if and only if those subsets have the same members. This assumes the Axiom of Extensionality for sets according to which, any set A and set B are identical if and only if the members of A are exactly the same as the members of B. Perhaps that is the reason for describing this view as an extensional account of relations.

If someone had reservations about identifying relations with sets (as (1) and (2) require), we can always advert to a theory which does not identify relations with sets, but which describes a close connection between them. Accordingly, we could have a *non-reductive theory* of extensional relations which would require that

(1′) *Every* n-place relation on S is associated with a subset of $S^n$.

and

(2′) *Every* subset of $S^n$ is associated with an n-place relation on S.
(3′) If the relations R and U are the same, then their associated sets are identical.

If we asserted

(4′) If R and U are relations and their associated sets are the same, then R is identical with U,

the converse of (3′), then we would have a theory that says that any relation is associated with a set such that one relation is identical with another if and only if their associated subsets are identical. Assuming the Axiom of Extensionality for sets, this too would be a theory of relations that could be called an extensional theory of relations. Its one virtue is that it would not identify relations with sets, if that were thought a virtue. For those who set store by not reducing relations to sets, the primed theory shows that you can have an account of relations as extensional, without the reduction. But the cost is not free: an account of the notion of a relation's *corresponding* to a set now has to be explained, and of course one also has to explain what these relations are supposed to be. If you deliberately deny (4′):

(4″) There are two relations R and U, such that their associated sets are the same.

A theory consisting of (1′) – (3′), and (4″) would be a non-extensional theory of relations.

We could also indicate how a non-extensional theory of properties might go, by essentially using the theory for relations, with properties as a special case. We could start by thinking of extensional properties as properties that are identified with the subsets of a given set S.

People who take properties seriously do not do that. Nevertheless it is common in discussions of models or interpretations in logic to say that to each predicate letter of a logical system, an interpretation assigns a property –i.e. a subset of the domain or set S of the model. So properties, according to this usage, would be sets. Again, to avoid this reduction, we could think of an extensional theory of properties which would require that

(1) *Every* property on S is a subset of S, and
(2) *Every* subset of S is a property on S.

Again, to avoid this straight identification of properties with sets, we can modify the theory as we did for relations. We would require that

(1∗) *Every* property on S is associated with a subset of S, and
(2∗) *Every* subset of S is associated with a property on S.
(3∗) If the properties P and Q are the same, then their associated sets are identical.

If we were to assert

(4∗) If P and Q are properties and their associated sets are the same, then P is identical with Q,

the converse of (3∗), then we would have an extensional theory of properties, in which properties are not identified with sets, not even with the sets of those things that have them). The costs of doing this are the same as those mentioned in connection with the parallel theory of relations described above – the need for some rigorous account of "association". Now if we deliberately deny (4∗) and adopt

(4∗∗) There are properties P and Q, such that their associated sets are the same, but P is not identical with Q.,

then a theory which consists of (1∗) – (3∗), and (4∗∗) would be a non-extensional theory of properties.

Here is a possible example: Consider the committees $C_1$, $C_2$, ..., $C_n$, of some organization. Every committee may have some members, though sometimes none. To that committee there is *associated* the set of its members. It is quite acceptable to have different committees that have exactly the same associated set of members. If we take "being a member of a committee" as a property (cf. Dretske 1977), then the practice according to which different committees can have exactly the same members, would support condition (4**). The committee structure would have to be modified a bit to satisfy condition (2*), but the rest seems to be fine.

**(a2) The Nonextensional Version** Unlike the extensional view of relations, there is no canonical version of the nonextensional view. It is best to start with a simple example, due to Frank Ramsey (slightly modified) that focuses on his modification of Russell's notion of a propositional function. Similar remarks are made about relations and functions, and in the case of relations, Ramsey called his modification the notion of a **relation in extension.**

Ramsey, following Russell, regards a propositional function as a mapping from individuals to atomic propositions. In what follows we shall ignore the restriction that all the values of these functions are atomic. That restriction may be important in contexts where the issue is whether propositional functions are universals. Where it makes a difference in what follows, we shall say whether the propositions are atomic or any truth-function of atomic propositions.

On the Russellian view (In the *Principia* and elsewhere), an additional constraint is placed on propositional functions, by imposing an additional requirement on all the values of the propositional function. For example if P is a propositional function such that

P(Socrates) is the proposition that Socrates is wise, and
P(Plato) is the proposition that Plato is wise (and so on),

(where for the sake of the example, it is not necessary to specify what the values of the function are for other objects). The disagreement with Russell does not concern the mapping which assigns to a, the proposition that a is wise, to b the proposition that b is wise, and so forth. Such a mapping is clearly a propositional function. The disagreement raised by Ramsey emerges over a function Q for which

Q(Socrates) is the proposition that Queen Ann is dead, and
Q(Plato) is the proposition that Einstein is a great man. [1990, (215)]

Q certainly is a mapping from individuals to propositions (it does not matter here what account of propositions is used). On Russell's view, however,

Q is not a propositional function. The reason for the difference seems to be the vague idea that the proposition, Q(Socrates), does not say the same thing of Socrates, that Q(Plato) says of Plato. Ramsey made it central to Russellian propositional functions that if P is a propositional function, then for any arguments a and b, the values, P(a) and P(b) for them are such that P(a) says the same thing of a that P (b) says of b.

Thus predicative propositional functions are different from functions construed extensionally. Their values cannot be arbitrary propositions; their values are constrained. As Ramsey stated it:

> We call such functions 'predicative' because they correspond, as nearly as a precise notion can to a vague one, to the idea that Pa predicates the same thing of a as Pb does of b. They include all the propositional functions of *Principia Mathematica*, including identity as there defined.

And the real point of this observation immediately follows:

> It is obvious, however, that we ought not to define identity in this way as agreement in respect of all predicative functions, because two things can clearly agree as regards all atomic functions and therefore as regards all predicative functions, and yet they are two things and not as the proposed definition of identity would involve, one thing.[2] [1990, *Foundations of Mathematics*, 212–213)

Ramsey's target in "Foundations of Mathematics" (FM), was Russell. Here are two passages from Russell and Whitehead's *Principia Mathematica* that sustain Ramsey's objection (the actual passages are not cited by Ramsey in his article). The first passage asserts

> Given any elementary proposition which contains a part of which an individual a is a constituent, the other propositions can be obtained by replacing a by other individuals in succession. We thus obtain a certain assemblage of elementary propositions. We may call the original proposition ϕa, and then the propositional function obtained by putting a variable x in the place of a will be called ϕx. Thus ϕx is a function of which the argument is x and the values are elementary propositions. The essential use of "ϕx" is that it collects together a certain set of propositions, namely all those that are its values with different arguments.[3] (**PM to ∗56**, 1962, p. xx).

This makes it evident what the propositional functions are supposed to be, and how they are formed with aid of essentially atomic propositions by the replacement of the name in an atomic proposition in which it occurs, by a variable. The restriction to the use of atomic propositions can, as we have said be relaxed. But the passage, although it ignores the distinction between what we would call an open sentence, and a function, nevertheless makes it clear what the propositional functions are supposed to be, and explains why Ramsey called this kind of propositional function predicative.

Russell's use of propositional functions, from a present perspective, was all over the place. He thought of them as linguistic (what remained after a referring term is struck from a sentence) –what would now be called an open sentence, and he also thought of them as a kind of function that assigned to each object a, the proposition (or sentence) obtained by substitution of the name for the variable (or filling in the place where the name had been struck) in an expression which is an open sentence. He also thought that propositional functions might be systematically employed rather than classes or properties –though some care has to be made in these ascriptions as his views developed over time.

Although we have focused on Ramsey's version of propositional functions in the 1920s, there is nevertheless a close connection between predicative propositional functions and properties, as noted by Richard Braithwaite in his fellowship essay A Dissertation on Causality (1923):

[2] F.P. Ramsey, Philosophical papers. D.H.Mellor (ed.).

[3] Principia Mathematica to ∗56.Cambridge Mathematical Library Pb, 1997.

> I am not using "property" in the most general sense as "Propositional function of one variable", but in the sense in which (as I explained above) Mr Keynes uses it. (102)[4]

The connection can be seen by the following admittedly somewhat inexact two observations: (1) to every property we can associate a (predicative) propositional function – the one which assigns to any object a, the proposition which states that a has that property, and (2) to any (predicative) propositional function we can associate a property P as follows: since all the propositions P(a), P(b), P(c), ... which are the values of the predicative propositional function all predicate the same thing of a, b, c, ..., let what is predicated in all these cases be the property associated with the (predicative) propositional function. These considerations only support the view that there is a connection between predicative propositional functions and properties. They do not support the reduction of one to the other.

The second passage from Russell and Whitehead's Principia exploits the difference between variations involving the singular term "a" in a proposition, and variations on the remainder with the term a removed.

> The difference between a function of an individual, and a function of an elementary function of individuals is that, in the former, the passage from one value to another is effected by making the same statement about a different individual, while in the latter it is effected by making a different statement about the same individual. Thus the passage from "Socrates is mortal" to "Plato is mortal" is a passage from f!x to f!y, but the passage from "Socrates is mortal" to "Socrates is wise" is a passage from $\phi$!a to $\psi$!a. Functional variation is involved in such a proposition as: "Napoleon had all the characteristics of a great general." (**PM to ∗56**, 1962, p. xxix).

From this second passage it is evident that Russell thinks of propositional functions as generated from the variation on the singular term "a" which occurs in a statement, and thinks that to each individual, the propositional function assigns a statement which in every case says the same thing, only about a different individual. Russell's distinction between the two kinds of variation on propositional functions will be familiar to those who know Ramsey's paper on Universals. By the first variation the result is the same statement (sic) about a different individual, and the second kind of variation results in a different statement about the same individual. Ramsey's important challenge to the logical significance of this distinction cannot be addressed here.

Although in the passage cited, Ramsey uses the notion of "same predication", he does, in a later passage of the same paper use a different description that avoids the use of any notion of predication, in favor of the notion of what a proposition says about an individual.

> It seems therefore, that we need to introduce non-predicative propositional functions. How is this to be done? The only practicable way is to do it as radically and drastically as possible; to drop altogether the notion that $\varphi$a says about a what $\varphi$b says about b; to treat propositional functions like mathematical functions [sic], that is, extensionalize them completely. Indeed it is clear that, mathematical functions being derived from propositional, we shall get an

[4]This is an unpublished typescript which was Braithwaite's Fellowship Dissertation. It is on deposit in the King's College Archives, Cambridge University, England.

adequately extensional account of the former only by taking a completely extensional view of the latter. [Mellor 1990, in FM, (215)]

Ramsey offered two ways of understanding Russell's constraint on the values of propositional functions.

According to one version, the first proposition φ(a) predicates the same thing of a that the second proposition φ(b) predicates of b. According to the second version φ(a) says the same thing of a, that φ(b) says of b. The "same saying" version seems more general than the one that appeals to predication. Ramsey makes it fairly clear that if there are no predicates (Ramsey sometimes says "actual" or "real" predicate or formula) of the appropriate language under study) which are uniformly predicated, then there is no propositional function. The same-saying version seems less tied to what is expressible in a particular language. It is especially interesting to note that Ramsey connects up the requirement that there be *actual* predicates that can serve to define a propositional function with the related issue (which we shall describe below) of whether there has to be an actual or real formula for there to be a function of a real variable (Mellor 1990, FM, 178).

This indicates, to be sure, somewhat allusively, that Ramsey was aware of the fierce mathematical debate that raged over the proper concept of function (fully extensional) that would serve the needs of mathematics and mathematical physics. Ramsey proposed a parallel revision of the concept of propositional functions that was used in the first edition of *Principia Mathematica*.

There is the strong belief that Ramsey knew of controversy over the shift from the notion of an nonextensional account of them to an extensional account of them and that he wanted, in a parallel way, to replace Russell's use in the *Principia* of the older notion of what he called predicative (nonextensional) propositional functions to one that was aligned with the more contemporary notion of function advocated by contemporary mathematicians and physicists.

This belief is supported by the discovery by Professor Michael Potter that Ramsey possessed a copy of Hausdorff's *Mengenlehre*. Potter discovered a copy in a London bookstore, signed by Ramsey. Hausdorff explicitly mentions Dirichlet's general concept of a function, and said "The conception of a function is as fundamental and primitive as the concept of a set. A functional relation is formed from *pairs of elements* just as a set is formed from *individual elements*" and Hausdorff thought that this concept allows this fundamental concept its full scope and generality.[5]

Conceptually, both revisions call for a fully extensional account of functions, mathematical and logical. The failure to attend to the conceptual change in the concept of a function lies we believe at the heart of Armstrong's flawed account of scientific laws. We shall explain this in the following chapter.

[5]F. Hausdorff, *Mengenlehre*, First German ed., 1914. Potter says that unfortunately there are no underlined or marked passages in it. The quotation above is from the preface to the third English edition, *Set Theory*, Chelsia Publishing Company, Tr. J.R. Aumann et al.(15–17)

First however, let us return to Ramsey's proposed modification of *Principia Mathematica*. It involved the replacement of Russell's use of *predicative propositional functions* (Ramsey's term for them) by their more general extensional cousins. Ramsey left it open as to whether functionals, that is, functions of functions could be taken as predicative. He thought that for technical reasons, there was no need to provide an extensional version of functionals, at least as far as the needs of Russell's account of arithmetic were concerned. This may be so, but as we shall see, in physics, the notion of a functional is central for the expression of significant laws. It is not obvious that their use can be bypassed. As we shall see later in the second part of this study, functionals play a significant role in contemporary mathematics and physics, and for that reason an extensional version of them is also needed.

We shall not try to distinguish Ramsey's two ways of explaining the Russellian constraint on propositional functions. Instead we shall explore the subject a bit further by adopting a more general way of describing the constraint on propositional functions in a way that includes both of the ways that Ramsey considered. The use of the more general but simpler version of the Russellian constraint will also enable us to see why the confinement to nonextensional relations and functions would leave the mathematical sciences severely impoverished in descriptive and explanatory power.

Russell's particular constraint in either of the two forms that Ramsey gave to it, suggests that there is an equivalence relation R among certain kinds of propositions. That is we can say that $\varphi(a)$ and $\varphi(b)$ stand in the relation R if and only if they say the same things about their respective arguments, or we can say that they stand in the relation U if and only if they predicate the same thing about their respective arguments. Vague as the claims may be, they suggest that in either case there is an equivalence relation on some set of propositions (I.e. a binary relation that is reflexive, symmetric, and transitive[6]).

Let us assume then that we have a mapping P∗, from some set of objects O to some set S of propositions, and that R is some equivalence relation on S. R is clearly a way of avoiding the difficult task of explaining the vague notion that P(a) says (or predicates) the same thing of a that P(b) does of b. We shall assume that there is some equivalence relation R on propositions that will serve to express the idea that some of the propositions of S are like each other, and some are not. It is clear that in each of the two ways of expressing Ramsey's notion of "saying the same thing" are meant to appeal to an equivalence relation that is different from the identity relation on S.

In what follows, we shall use asterisks to indicate nonextensional notions. If we specify a non-empty set S (of at least two propositions), together with an equivalence relation R on S other than the identity relation, then we can define a *nonextensional propositional function (of one argument) on S, with respect to R*, this way:

> Let X be a non-empty set of objects, S be a non-empty set of propositions or sentences, and R be an equivalence relation on S other than the identity relation. Then *P∗ is a non-extensional*

[6]That is R(x, x), If Rx, y) then R(y, x), and If R(x, y) and R(y, z), then R(x, z), for all x, y, and z.

> *propositional function mapping X to S,* with respect to the equivalence relation R, if and only if for all objects a,b, . . ., in X, all the propositions P∗(a), P∗(b),. . . belong to one equivalence class of R, on S.

This generalizes (a bit) the kind of propositional function that Ramsey called "predicative propositional function" (ppf) and which we have called *non-extensional.*

So for us, non-extensional propositional functions are always non-extensional with *respect to a specific equivalence relation R, other than identity,* on the set of propositions that objects are mapped to.

Extensionalized propositional functions are very different. For them, there is no requirement that there be an equivalence relation on the values of the function. The mapping is to any set of sentences. That is,

> An *extensionalized propositional function is a function from a non-empty set of objects X to any non-empty set of propositions or sentences S* such that (1) $\mathfrak{R}$ is the set of all pairs, <o, s >, where o is any member of X, and s is any member of S, and (2) for any <x,y> and <x,z> in $\mathfrak{R}$, $y = z$. This covers the case of functions of a single argument.

It is easy enough to define the notion of a (nonextensional) function of several variables.

This covers the case of functions of a single argument. The notion of a (nonextensional) function of two variables can be defined this way: Let T∗ be mapping of pairs of objects <a,b> into a set of propositions S, and let R be an equivalence relation on S. Then

> T∗ is a nonextensional function of two variables on S if and only if it maps all pairs of objects <a,b> to S (propositions), to the same equivalence class.

Thus, T∗(a,b) and T∗(c,d) are alike in the sense that they belong in the same equivalence class -that is, they stand to each other in the relation R. The case for non-extensional functions of more than two variables is the obvious generalization.

We shall see below how an earlier parallel shift from the use of non-extensional functions to the use of extensional functions took place in mathematics and physics. For the present however, we shall focus on Ramsey's proposal to endorse a similar shift from the use of non-extensional to extensional propositional functions in the *Principia Mathematica* of Russell and Whitehead. Ramsey thought that the mathematical notion of identity could not be defined the way that Russell thought it could, using non-extensional propositional functions. That is, define "$x = y$" as $(\phi)(\phi x \equiv \phi y)$, where quantification was over all nonextensional propositional functions.

The result would be that standard arithmetical concepts and theorems could neither be expressed nor proved. This would of course be a serious criticism, in addition to other serious ones that Ramsey raised. After all what was at stake was Russell and Whitehead's aim to capture much of traditional mathematics in their *Principia.* That goal could not be attained if it turned out that a key notion for arithmetic such as identity, could not be provided by their account of it.

In the remainder of this chapter we want to discuss these limitations: first in arithmetic, and second in physical science, that comes with the exclusive use of non-extensional propositional functions. First, to the problem of identity.

So far we have discussed predicative (nonextensional propositional functions (ppfs) (those which we have called nonextensional). It is interesting to note how advocates of ppfs take account of the identity relation x = y.

Is the mapping of pairs of objects, <a,b>, to the proposition a = b, a nonextensional relation or not? Some philosophers, (Wittgenstein for example) did not believe that for two names, or even for the same name, that "a = b" has any sense; there is therefore no problem of defining it. Identity, he famously proclaimed, can be *shown*, presumably by a suitable notation, but it cannot be *said*. And if it cannot be said, then it cannot be defined either. If there was no definition for identity. It might then be taken as a primitive notion. In that case however, there would not be a problem of defining it.

For Russell, however, there was a problem, and a special one, because in *Principia Mathematica* he defined "x = y" as (ϕ)(ϕx ≡ ϕy), where the quantification is over all predicative (nonextensional) propositional functions "ϕx".

If I∗(x,y) = (ϕ)(ϕx ≡ ϕy) is understood as quantifying over all predicative (nonextensional) functions ϕ, then it looks like I∗(x,y) is also nonextensional. There's no doubt that it is reflexive, symmetric, and transitive (as proved by Russell in PM 13). That would show that I∗ is an equivalence relation; not necessarily the identity relation. It would still follow from Russell's definition that I∗(x,y) is a tautology when x and y have the same "meaning" (reference?), and is a contradiction when they are different in meaning. But that goes no way towards showing that identity has been defined, rather than some surrogate. Here is Ramsey's take on this issue:

> It is obvious, however, that we ought not to define identity in this way as agreement in respect of all predicative functions, because two things can clearly agree as regards all atomic functions and therefore as regards all predicative functions, and yet they are two things and not, as the proposed definitions would involve, one thing. . . .
>
> If we are to preserve at all the ordinary form of mathematics, it looks as if some extension must be made in the notion of a propositional function, so as to take in other classes as well. Such an extension is desirable on other grounds, because many things which would naturally be regarded as propositional functions can be shown not be predicative functions. [Mellor 1990, 213]

Continuing on, he stressed that a new notion of propositional function is needed, by providing a vivid example of the new kind of propositional function that was not a predicative propositional function:

> So in addition to the previously defined concept of a predicative function, which we shall still require for certain purposes, we define, or rather explain, for in our system it must be taken as indefinable, the new concept of a propositional function in extension. Such a new function of one individual results from any one-many relation in extension between propositions and individuals; that is to say, a correlation, practicable or impracticable, which to

every individual associates a unique proposition, the individual being the argument to the function, the proposition its value.[7]

Thus the propositional function φ whose values for Socrates and Plato, φ(Socrates) and φ(Plato), respectively, may be Queen Anne is dead, Einstein is a great man; . . ." would certainly not be a Russellian propositional function.[8]

The point to be stressed is that this new function is *any* one-many relation in extension. There is no special requirement on the values of the function that requires them to say the same thing for each of the different arguments of the function.

Ramsey thought that the relation of "not other than" is not predicative. His own proof of that fact is rather complex, and here we provide what we hope is a simpler proof of that result: Consider the relation "$x \neq y$". Let b be an object, and let a be any object other than b. Consider the propositional function "$x \neq b$" ("x is other than b"). It cannot be predicative. For consider the two values this function has for the arguments a and b respectively: $a \neq b$, and $b \neq b$. The first is true, and the second inconsistent. So the first doesn't say of a, the same thing that the second says of b. Thus "$x \neq b$" isn't a predicative propositional function, and (Ramsey believes) consequently neither is "$x = y$", the negation of x is other than y.

An argument for the same conclusion can be given using the more general description we have provided for predicative propositional functions. Consider the relation "$x \neq y$". Let b be any object. Then the propositional function "$x \neq b$" is not a predicative propositional function. If it were, then the propositions "$a \neq b$" and "$b \neq b$" would have to be in the same equivalence class. That however, could be ruled out by two plausible conditions on equivalence classes according to which (1) no two sentences can be in the same equivalence class if their conjunction is contradictory, and (2) if two sentences are in some equivalence class, then so too are their negations. It follows from (1) that "$a \neq b$" and "$b \neq b$" cannot be in the same equivalence class, and it follows from (2) that "$a = b$" and "$b = b$" aren't in the same equivalence class. And so, "$x \neq y$" is not a predicative propositional function. And if it is not a predicative (nonextensional) propositional function, neither is $x = y$, its negation, a predicative propositional function. So the identity relation is not available if one insists on using only nonextensional functions. Identity is beyond the predicative pale, if one restricts oneself to predicative propositional functions.

We should note in passing that Wittgenstein seems to have been very critical of Ramsey's proposal for the adoption of extensionalized propositional functions, and there was a brief correspondence of sorts between them, which despite P. Sullivan's remarkable discussion of the issue, leaves the point of Wittgenstein's complaint as opaque as ever.[9]

---

[7]Mellor 1990, (215).

[8][Mellor 1990, (215)].

[9]There is a letter of Wittgenstein to Ramsey critical of Ramsey' paper "Foundations of Mathematics (reprinted in Mellor, [1990]., and Wittgenstein asks Ramsey to convey his response to M. Schlick. Part of Ramsey's response is conveyed by Schlick to Wittgenstein. These letters and responses are reprinted in *Wittgenstein and the Vienna Circle*, Blackwell, Oxford 1984. Ramsey then drafted two

### 5.1.1 *Nonextensionality and Scientific Laws*

If our account of the nonextensional functions is near the mark, then there are some simple issues that indicate a conflict with mathematical practice. For example, one familiar way of describing a function is as a relation R(x,y) such that if R(x,y) and R(x,y*), then y = y*. Obviously something will be needed to explain functions as a special kind of relation, and it's not evident how this can be done without identity entering somewhere into the account. Another simple problem is to describe sets of finite cardinality. The usual way of describing a set as having two elements is to say that some non-identical x and y belong to the set, and every member of the set is either identical to x or to y. Again it's not clear how this is to be done without identity. These are but a small fraction of the examples where identity figures prominently.

We shall set these particular mathematical examples to one side, and consider in addition some mathematical and physical examples that proved to be important for the sciences. We want to concentrate on the havoc that the restriction to nonextensional relations and functions can have on the physical sciences, and nearly did.

The point that we shall try to establish is that the demand that physics (for example) restrict its practice to nonextensional functions would eliminate many of the succeses physics has had. We can illustrate that point with the help of two crucial episodes in the history of science when it became evident that the exclusive use of nonextensional functions stood in the way of a proper solution to some key scientific problems.

The first episode concerns a controversy between L.Euler and J.D'Alembert[10] over the general solution to the problem of vibrating strings, and a second, later episode, that concerned the scope of the domain of validity of the Fourier analysis of functions, and the theory of heat flow. There have been many ways of describing these critical junctures in the history of science, but the salient issue for the present purpose is that it concerns a difference over whether an older way of thinking about functions, – a nonextensional one, was to give way to an extensional account.

One immediate problem facing any discussion of the use of nonextensional functions in mathematical physics, is the lack of a clear idea of what this means. In our earlier discussion of nonextensional propositional relations and functions, the issue was somewhat easier to describe. In the propositional case we used equivalence relations on "propositions" – the values of propositional functions, and placed a few

responses to Wittgenstein directly. It's not clear whether they were ever sent. They are reprinted in M.C. Galavotti, *Notes on Philosophy, Probability and Mathematics*, Bibliopolis, 1991, pp. 337–346). Also cf. the noteworthy paper of P.M. Sullivan, "Wittgenstein on "The Foundations of Mathematics", June 1927, in Theoria 61(2), 1995, pp. 105–42.

[10]Admirably discussed in I. Grattan-Guinness, *The Development Of the Foundations of Mathematical Analysis From Euler to Riemann*, MIT Press (1970), pp. 2–12), and I. Kleiner, "Evolution of the Function Concept: A Brief Survey, *College Mathematics Journal*, 20 (1989), pp. 282–399.

reasonable conditions on those propositions so as to reflect the intuitive but vague idea that the proposition "P(a)" says the same thing of a, that the proposition "P(b)" says of b.

In the cases we now wish to consider, the values of functions are taken to be (in the simple cases) real numbers rather than propositions. It is not clear how the propositional function case can be applied here. The kind of constraints on the values of propositional functions that Ramsey mentioned, were concerned with same-saying, or predicating the same thing of two different objects. None of those concepts have anything to do with numbers, vectors, tensors or other mathematical objects – i.e. the values of the usual functions deployed in mathematical physics. Numbers (unless they are part of a coding scheme) don't say anything, and so don't say or predicate the same things as other numbers.

Nevertheless, there is a similarity we think, between the way that Ramsey characterized Russell's notion of a propositional function, and the older notion of a function employed by most physicists at least) from the late Eighteenth century to pretty much the early part of the twentieth. Here is how that early concept of function could be characterized:

> A function is a mapping f, from a set of objects O to the real numbers R, together with an algebraic or transcendental formula or condition A[x], such that all the values of the function f satisfy that condition. I.e. A[x] = f(x) for all x in some finite interval I over which the function is defined.

It also assumed that the algebraic and transcendental formulas which characterize the values of functions are definitive – i.e. different formulas yield different functions.

There are three ways in which this early equational concept of a mathematical function, and the Russellian concept of a propositional function are similar. First, the specific values for the older concept of mathematical function requires that f(a) for the argument a, is given by A[a], and the value for f(b) is given by A[b]. These values are related by an equivalence relation in that they are the results of different substitutions in the one formula A[x].[11] Likewise, the values of Russellian propositonal functions (such as "x is wise") are alike in that they are various substitution instances of "x is wise"(Socrates is wise, Plato is wise, etc. etc.). In the case of the Russellian mathematical functions and in the case of the later extensional propositional functions, the values are alike in that in each of the cases, there are equivalence relations, and the values all fall within one equivalence class of those equivalence relations.

There is another, and perhaps the most important way in which Russell's propositional functions and the nonextensional mathematical functions match up. It is the expressive weakness that results with the exclusive restriction to the non-extensional functions. In the case of the propositional functions there was, as Ramsey argued, no definition of identity available in arithmetic.

---

[11]That is, aRb if and only if A[a] & A[b].

We shall now argue that the restriction to the use of only non- extensional functions, results in the failure to have solutions to important problems in the physical sciences.

Here is a telling passage from the mathematician Cauchy written in 1844, reflecting back on the work of Euler and Lagrange almost a 100 years earlier, and compactly described by I. Grattan-Guiness.

> In the works of Euler and Lagrange, a function is called *continuous* or *discontinuous*, according as the diverse values of that function, corresponding to the diverse values of the variable ... are or are not produced by one and the same equation .... Nevertheless the definition that we have just recalled is far from offering mathematical precision; for the analytical laws to which functions can be subjected are generally expressed by algebraic or transcendental formulae [that is, by the Eulerian range of algebraic expressions], and it can happen that various formulae represent, for certain values of a variable x, the same function: then for other values of x, different functions.[12]

As Grattan-Guinness noted, Cauchy was not accurate on the historical account of continuity that was in use, prior to his own definition. However, Cauchy was right about the earlier requirement on the special way in which functions were supposed to be related to their arguments. The values of a function were tied to equations of the sort: $y = A[x]$, where the right hand side of the equation is an algebraic or transcendental expression, and for any value of the variable x, say a, the value of the function is given by A[a]. The now standard extensional view requires that a function is given by *any* collection of ordered pairs (x. y), (subject of course to the condition that if (x,y) and (x,y′) then $y = y'$). The older account requires that not just any set of ordered pairs is associated with a function. It is only to those sets of ordered pairs $<\alpha,\beta>$, such that the equation $\beta = A[\alpha]$ holds, for some algebraic or transcendental expression A[x], that there is associated some unique function $f_{A[x]}$, say, such that $f_{A[x]}(z) = A[z]$, for all z. Let us call this the *equational concept of function.* It follows from this equational view that you cannot have functions whose values are given by different algebraic or transcendental formulas for different values of the variable in the domain. It is this consequence that concerned Cauchy, because he had been worried about certain integral representations of functions, for which the value of the function would be given by different formulas for different members of the domain.[13]

Essentially, the kind of example that Cauchy used is one in which a function is defined by cases, in which the function is specified by some expression or formula over part of an interval, and a *different* expression is used over the rest of the interval. There is no one formula that is used uniformly over the entire interval. The equational view, a view prior to Euler, embodied a critical condition which can be succinctly summed up this way:

[12]*The Development of the Foundations of Mathematical Analysis from Euler to Riemann*, MIT, Cambridge, 1970, p. 50.

[13]I. Grattan-Guinness, p. 51.

> One of the basic features of the old theory was that the algebraic expressions involved were understood to operate over the *whole* of their range of definability[14]

There is a bit of a problem in seeing why this is a problem. Why couldn't one have the function f defined over say the interval [0, a), and g defined over the remaining part of the unit interval, [a, 1], and combine f and g, by using the characteristic functions over those subintervals. That is, let h(x) be defined over the unit interval such that h(x) = f(x) $\delta_{([0,\ a)}$ (x) + g(x) $\delta_{[a,\ 1]}$ (x), where $\delta_{[0,\ a)}$ (x) is 1 for x in the interval [0, a), and 0 otherwise, and $\delta_{[a,\ 1]}$ (x) is 1 for x in the interval [a, 1], and 0 otherwise. The function h is clearly equal to the function f over the first subinterval, and is equal to the function g over the remaining interval. Moreover it looks as if there is a uniform way of obtaining the values of h over the interval. Just substitute for the variable 'x' in the expression provided by the definition.

There are two reasons why this obvious construction wouldn't be acceptable to the advocates of the older view. The first problem is one raised by Cauchy himself. There will be functions which fail to be what he calls "continuous", where he seems to mean what we now mean by "differentiable". He is certainly correct that in some cases, there will be non-differentiable functions if we depart from the older tradition. A very simple example consists of a v-function angling or "cornering" upwards above x = ½, (the apex), where there is no derivative (Fig. 5.1).

Differentiability is one thing that a restriction to a single algebraic or transcendental expression over the entire domain of the function guarantees. The second point is that although the proposed definition yields a function by our lights, it is not a function by theirs. There is the immediate problem that the function f sloping upward over the interval [0, ½) is not defined over the whole unit interval. Similar remarks hold for the function *g*. This defect can be remedied by replacing the functions f and g in our definition of h, by the functions f∗ and g∗ which are defined over the unit interval by the simple device of using the same algebraic expression for f∗ over the unit interval, that was used for f over the interval [0, ½). A similar extension can be defined to obtain g∗, using the algebraic expression used for g over the interval [½, 1]. The one remaining objection to piecing together functions f and g to obtain h is that the delta functions like $\delta_{[0,\ a)}$ (x), and $\delta[_{a,1]]}$ (x) which are needed, are not equational kinds of functions. There is for the characteristic function $\delta[_{[a,1]}$ (x) (for example) no algebraic expression C[x] which has the value 0 for the

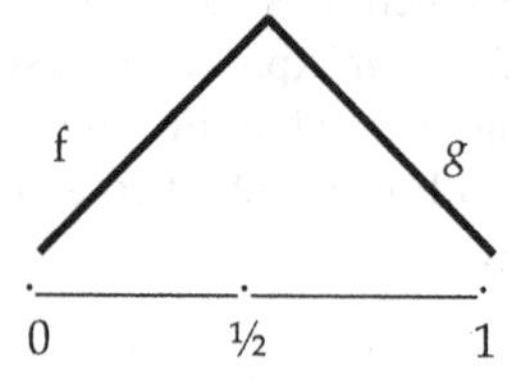


**Fig. 5.1** Example of an "angling" function

[14] I. Grattan-Guinness, p. 10.

infinitely many points in the interval [½, 1], and the value 1 for the infinitely many points in the remaining interval.

It was Euler, deeply indebted to this older view, who, in an exchange with D'Alembert, modified that view in a crucial way. Instead of insisting that functions be restricted in the old way, by having the same equation with a single algebraic or transcendental formula over its whole domain, he wanted to now include new functions which would be governed by one equation over one part of the domain, and another equation (with a different algebraic or transcendental formula over other parts of the domain.[15] Thus one could now have functions that are pieced together along an interval.

For example, Euler considered a function which was constant over the interval [0, a) but a cosine function over (a, l] which might be a solution of the vibrating string problem when the string is plucked at the point a. This new class of functions with their "corners" (as he termed them), would raise all sorts of difficulties for differentiation (Euler's notion of "continuity"), but they would allow him to give a general solution to the problem of the vibrating string. D'Alembert would have none of it, though Euler, in correspondence with him took a very positive note by affirming that

> Considering such functions as are subject to no law of continuity opens to us a wholly new range of analysis.[16]

At stake here was an adequate solution to a physical problem like the vibrating string. There were several competing solutions all claiming to cover the general case. It's fairly clear that no one algebraic or transcendental formula would do for solutions that would include the case for arbitrarily long or even infinitely long intervals.

The story of the ever-widening account of functions took a huge step when the idea arose of associating functions with infinite series. This was a major step in trying to overcome the limitation of using functions, each with one expression, substitutions in which would yield all the values for the function. Expressions would no longer be limited to those which were algebraic or transcendental, but would correspond to the values of infinite series for different arguments in a domain.

There were several prominent series that were considered, but it is Fourier's series (which arose in connection with his studies on heat diffusion) that had such a huge impact on the development of mathematical analysis and physics. Such a series is of course not an algebraic formula – not without extending the notion of a formula, so that the equational view of functions is out of its depth. But of course, even allowing an extended notion of formula so as to include equations involving infinite series, it still is not true that for every function, there is an equation between it and its Fourier series. More exactly, the idea is that a Fourier series associated with a function f, is one that involves only trigonometric terms with certain coefficients. Those coefficients are taken to be certain integrals involving f (over the appropriate interval). The

[15] I. Grattan-Guinness, pp. 50–51.

[16] I. Grattan-Guinness, p. 6.

series may or may not converge to the function f, so that the equation of a function f with its Fourier series is just incorrect: not every function is representable by a Fourier series. Counter-examples provided by Dirichlet and others made that clear. In fact there are continuous functions which are not representable as Fourier series. There are also celebrated theorems (a famous one due to Dirichlet) that specify the special conditions under which a function is identical to its Fourier series. It is clear then, that using just Fourier series will fall short of the class of extensional functions. Consequently, the extended equational view which went beyond the use of algebraic and transcendental expressions for obtaining the values of functions, to include the use of values that are obtained by the use of the Fourier series, could be seen as tenable only under certain conditions. In particular, the result of restricting functions to only those which converged to their Fourier series would seriously diminish the resources for the expression of general solutions to key problems in physics.

Let us return then to the equational view of functions. Is it a non-extensional view? Clearly so. Consider just the simple case of all the functions mapping the set of natural numbers to itself. Let f be the Dirichlet function which is the characteristic function of the rational numbers. If f(x) is c then x is a rational number and if it is d (diferent from c) then x is irrational. But there is no algebraic formula A[x] such that f(x) = A[x]. For then A[x] would have to be c on all the rationals and d on the irrationals. And that is impossible.[17]

The conclusion we draw from these reflections on an episode in the history of mathematics is that the restriction to non-extensional (equational) functions in physical theory would result in there being no solution sufficient for the problem of vibrating strings, or for the distribution of heat in a metallic bars. That is, a critical loss of descriptive and explanatory power.

Thus far we have considered two examples of functions that are non-extensional: the equational functions common in mid-Nineteenth century mathematics, with its restriction of values of functions to the values of a specific algebraic or transcendental formula, and the later use of Russellian propositional functions, where the values of those functions are restricted to the set of instances of a specific predicate.

We think that D. Armstrong's views on laws, which we will focus on in the following chapter, is vulnerable to the same kind of defect that was lodged against the older non-extensional concept of function in physics. In the Russellian case, the widespread use of non-extensional functions resulted in the impossibility of describing some key concepts in elementary arithmetic. In the case of physics, the exclusive use of functions that were non-extensional resulted in a descriptive inadequacy that made it impossible to describe or explain certain laws in physics.

In the next Chap. 6, *Armstrong's account of laws; identity lost, regained, and lost again.*), we shall show that even with some friendly amendments to Armstrong's

[17]Here and above, we follow the usual definition of "algebraic function" according to which u = f(x,y,...,z) is an algebraic function if and only if there is a polynomial F, such that F(x,y,..., z,u) = 0. Cf for example, R, Courant, Differential and Integral Calculus, vol.I, Interscience, N. Y.1937, tr.by E.J. McShane, p. 485.

account of laws (including a nice theory of predication due to John Carroll that we will add to Armstrong's theory of universals), the resultant theory cannot describe laws that rely upon the use of identity, so that all laws that relate physical magnitudes are beyond the reach of his account. Essentially, Armstrong's theory has talked itself out of the very subject matter that was the target of his initial inquiry.

# Chapter 6
# D. Armstrong's Account of Laws: Identity Lost, Regained, and Lost Again

## 6.1 Advantages and Disadvantages of Armstrong's Almost Exclusive Dependence on Universals in His Account of Laws

We have seen that Dretske explicitly required that laws related items like F-ness and G-ness, which would normally be understood as a reference to universals. However he also referred to laws as relating physical magnitudes. If all physical magnitudes, including refractive indices in particular, count, according to Dretske, as universals, then it seems to me that universals so understood, are in tremendous abundance, and set no limits for the expressive power of physical theories, beyond those set by the use of ordinary mathematical analysis. We noted in Chap. 4, there isn't any problem with those universals used by Dretske, *if* they are physical magnitudes.

However, there are deep problems with the well-known proposal of David Armstrong's account of scientific laws. They concern his account of laws as relations of universals, where, unlike Dretske, he does not think that universals (according to his account of them), are physical magnitudes.[1] That raises deep issues that we shall focus on. I don't, in any way, want to diminish the importance of other criticisms of Armstrong's account of laws that are already voiced in the literature, but the problems we think that are decisive against his account of laws all trace to his special use of universals.

The first problem concerns his discounting of certain standard laws as laws. The most obvious case is Newton's first law of motion, the Law of Inertia. It states roughly that

(1) If there are no forces acting on a body, then that body is not accelerating (it will continue with constant velocity (either zero ("at rest") or "in motion" with a constant non-zero velocity).

[1]Cf, D. Armstrong, *What is a Law of Naure*, Cambridge University Press, 1983, p. 111, fn.1.



A. Koslow, *Laws and Explanations: Theories and Modal Possibilities*, Synthese Library 410, https://doi.org/10.1007/978-3-030-18846-7_6

It has proven difficult for Armstrong to credit (1) as a law, because for him all laws relate universals, and all universals, on his account of them, are instantiated. Assuming however, that Newton's Law of Gravitation holds, there is never a time at which there are no forces acting on a body. Consequently, the antecedent condition is empty.

This is a case of what are called laws with vacuous antecedents, and Armstrong doesn't accept that conditionals with vacuous antecedents, are laws. It's his particular requirement that all universals are instantiated that stands in the way.

One can get around the problem of vacuous antecedents, however, by considering the contraposition of the First Law:

(2) If a body is accelerating, then there is some force acting on it.

Both of these sentences, (1), and (2), are generally regarded as equivalent expressions of the Law of Inertia. For example, as we noted earlier, J.C. Maxwelll as well as R. Feynman clearly take them to be equivalent.[2] Consequently, there is an equivalent way of expressing the First Law, where the antecedent is not vacuous, since there are accelerating bodies. So, given the two equivalent formulations of the law, (1), and (2), the first fails to be a law because the antecedent universal is uninstantiated, but (2) passes muster since the antecedent universal is instantiated. We assume for the sake of the example, that Armstrong's assumption that being accelerated, and being force-free are universals. It's a moot assumption, but let it stand, for there is more serious trouble ahead.

There is a second problem that is raised by consideration of the contrapositive of the First Law. If Newton's first law is represented as a conditional with the universal F as antecedent and the universal G as consequent, then it makes no sense to consider the equivalent contrapositive with the negation of G as antecedent, and the negation of F as consequent. The reason is that on Armstrong's view of universals, if U is any universal, then there is no universal which is its negation. Thus, even if we relaxed Armstrong's requirement that all universals are instantiated, so that (2) was a law relating universals one would not have the equivalence of (1) and (2), because (1) would involve the negation of universals, and so (1) would not be a law.

The equivalence of the two ways of expressing the Law of Inertia is beyond the reach of any theory that requires laws to be relations between universals. Moreover, this argument also covers not only the law of inertia, but any conditional law with universals for antecedents and consequents.

It could be objected that this untoward result holds on the assumption that Armstrong's claim that N(F,G) ("F necessitates G") as the uniform way for representing laws, does not commit him to the view that N(F,G) itself is a

[2]Cf. Chap. 2.

conditional. True enough. And if it is not a conditional, then there is no contrapositive. However Armstrong is committed to the Modus Ponens condition (MP), according to which[3]:

$$(\mathbf{MP})\ N(F, G) \Rightarrow (x)(\ Fx \supset Gx).$$

I.e., N(F,G) logically implies a universally generalized material conditional This condition (**MP**) was needed to support his claim that laws are stronger than what he termed the Humean representation of them as universally quantified material conditionals.[4] However the prospect for proving (**MP**) are dim for the following reasons:

(1) In the condition (**MP**), "F" and "G" are universals, but "Fx" and "Gx" are predicates. Armstrong hasn't said how "F" and "Fx" are related (similarly for "G" and "Gx".), if at all.
(2) Does he intend that we also have $N(\neg G, \neg F) \Rightarrow (x)(\neg Gx \supset \neg Fx)$? If we wanted to express the condition for the contrapositive of the law N(F, G)? That is impossible because on Armstrong's view of universals, there are no negations of universals which universal.[5]
(3) Armstrong can't mean by (**MP**) that $N(F,G) \Rightarrow (x)(F \supset G)$ is the proper formulation of (MP), for the reason that on Armstrong's view, there are no conditionals whose antecedents and consequents are universals.

Armstrong has a lot to say about universals, but almost nothing to say about predicates. Nevertheless we want to suggest that the matter is not entirely hopeless. We could introduce certain predicates that are associated with universals, and propose a replacement for the troubled (**MP**), that avoids these issues. Namely

$$(\mathbf{MP*})\ N(F, G) \Rightarrow (x)(F{*}x \supset G * x),$$

where F∗x and G∗x are predicates that are associated with the universals F and G, respectively. This can be done, as we will show, using an ingenious device suggested by John Carroll.

This argument, if correct, doesn't assume that "N(F,G)" is a conditional.[6]

---

[3]Armstrong, *What is a Law of Nature?* Cambridge University Press, 1983, p. 97, and (3). p. 156. While F and G are universals, "¬Fx" and "¬Gx" are the supposedly negative universals "x is not being F", and "x is not being G" respectively.

[4]Though it is assumed, the claim has rightly been subject to severe criticism by Bas Van Fraassen, and David Lewis to the effect that (MP) is not justified by Armstrong.

[5]Armstrong, *What is a Law of Nature*, p. 156,

[6]But it does assume that Armstrong has interposed a material conditional horseshoe) between two universals and that raises another problem since Armstrong's view on universals is that with the exception of conjunction, there are no negations, disjunctions, or conditionals between universals. If we read "F[x]" as "x's being F" – i.e. what Armstrong calls a state of affairs, then he has to explain what the conditional between states of affairs could possibly mean. We try below to remove that impediment by shifting to (**MP**∗).

There is no doubt that the modified (**MP**∗) would have an important role to play in an account of laws. Without it, it is hard to see how Armstrong's N(F,G)s can play any role in explanations and predictions. The impediment to accepting the modified (**MP**∗) is that Armstrong nowhere appeals to predicates and predication. We have very detailed accounts of his view of properties, and universals, but not of predicates.

I don't think that the matter is completely hopeless. There are two changes that would improve things somewhat. First, one could just adopt as part of the account of laws, Armstrong-style, the modified version of (**MP**), i.e. (**MP**∗), where "F∗x" and "G∗x" are assumed to be predicates that are associated with the universals F and G respectively.[7] This notion of predication that can be associated with respect to universals, can be spelled out neatly using an idea due to John Carroll:

If F is a universal, then for any particular x, and universal F define the an associated predicate F∗ as follows[8]:

$$(\mathbf{C}) F^*(x) \text{ if and only if } F[x] \ (x \textit{ instantiates } F).$$

That is, the *predicate* F∗x associated with the universal F holds of any particular x, if and only x instantiates the universal F. We do not assume that instances of predicates are the same thing as instantiations of universals. An item a satisfies the predicate "F (x)" if Fa, and that a is an instance of the predicate. But we say that a particular a instantiates the universal F.

We can therefore save some version of Armstrong's claims with the use of (MP∗), provided we assume the condition (**C**). I don't know whether (**C**) would have been acceptable to Armstrong. We proposed adding on (**C**) only for those universals that occur in the expression of laws. Nevertheless the addition does advance his account somewhat, in that it shows how, on his notion of laws, N(F,G) is stronger than the universally quantified conditional (x)(F∗x ⊃ G∗x).

That's one small step. However there is another problem, involving the identity relation, that arises as a consequence of these additions, and it is in my view the most serious one yet. We hasten to say that the difficulty depends upon adding (**C**), which adds a natural connection of universals that occur in laws with predicates and relations. It seems to me that it is a natural and helpful connection. Nevertheless it is strictly speaking not a problem that can be laid at Armstrong's door, unless one leaves that door slightly ajar.

---

[7]Yes, this by fiat. But since it is needed, and Armstrong does not have a proof, then it could be argued that assuming (**MP**∗) is a maneuver that is no different than what D. Lewis did when he needed to assume that the counterfactual conditional in his system VC satisfied Modus Ponens. He added it on. It was needed since it didn't follow from the other statements of VC.

[8]The insightful proposal of John Carroll for defining a predicate in terms of a universal is in *An Introduction to Metaphysics*, N. Markosian and J.W. Carroll, Cambridge University Press, 2010, p. 229. Binary relations such as R(xy) are connected by Carroll neatly this way "a Rs b if and only if a and b (in that order) instantiate R-ing", where 'R-ing'is understood to be a two-place relation that is a universal. The variable "x" in (**C**) is understood to range over the particulars of Armstrong's account of universals.

## 6.2 Universals, Predicates, and Propositional Functions

Armstrong avails himself of the notion of a particular a instantiating a universal F, calling a's instantiating of F a *state of affairs*, where the particular a and the universal F are the only constituents of that state of affairs. We will indicate such states of affairs by "F[a]" (using the square-bracket notation We can then define an analogue to Russell and Ramsey's notion of propositional function.

To any universal F, we shall define a *state-of-affairs function* (SAF), $P_F$, as a function which assigns to any particular a, a state of affairs, F[a] (a's instantiating F). I shall refer to such functions using upper case letters subscripted by universals, to differentiate them from the Russellian notion of a propositional function. Thus for some universal F, $P_F(a) = F[a]$, for all particulars a. The difference between the two accounts (Russell's and the one we now wish to add to Armstrong's account, lies only in using particulars rather than objects, and states of affairs rather than propositions. Those differences seem to me not to matter for what follows.

Consider then two states of affairs F[a] and F[b]. The former states that the particular a instantiates F and the latter states that the particular b instantiates F. That is, the first states the same thing about a, that the second states about b. These notions are defined for using one particular and one universal, but the generalization to states of affairs which may have several particulars and several universals as their only constituents is easily defined as well.

Let's define W to be the set of all states of affairs i.e. F[a], G[b], H[c], …, We then can then define an equivalence relation **E** on W as follows:

For any states of affairs, say F[a] and G[b}, define F(a) E G[b] if and only if F and G are the same universal.

Now, for any universal F, consider the state of affairs function $P_F$. The set of all values of this function all belong to an equivalence class of the relation E -namely {F[a], F[b], F[c]….}. That means that the (SAF) functions are nonextensional.[9] It follows that these functions are not sufficient to define an identity relation between particulars if one requires that

a and b are identical if and only if for all universals U, particular a instantiates U if and only if particular b instantiates U.
i.e. a = b if and only if U[a] if and only if U[b].

This is something that should worry any advocate of Armstrong's account of laws, because without an identity relation, it is impossible to express many scientific laws. The reason is that most of those laws involve functions, and functions are, by a standard account of them, relations that appeal to identity. That is, functions of one

[9]The additional requirement that the equivalence class of all the values of $P_F$ are a proper subset of W is guaranteed if there are at least two states of affairs in W that are not equivalent by the equivalence relation E.

variable (for example) are binary relations R(x,y), such that if R(x,y), and R(x,z), then y is identical with z. All physical magnitudes are functions in this sense.

There is also a second way in which identity enters into a discussion of laws. Some laws, as we have seen are expressed as equations between magnitudes. We have already mentioned Ohm's Law (Chap. 4) and the Boyle-Charles law that PV = rT (The product of the pressure and the volume of a gas is equal to r times its temperature), to mention two simple cases.

There doesn't seem to be a way to define an identity relation, if one is restricted to the use of universals and what can be defined with their aid – such as Armstrong-type states of affairs. How then, to make up for the expressive deficiency of such a restriction?

One possibility would be to just add the relation of identity to Armstrong's account of laws. That would certain help with the expression of those functions that are physical magnitudes, and help as well with the expression of those laws that are usually represented as equations relating physical magnitudes.

Unfortunately, supplementing his theory with the identity relation is not permissible for Armstrong. He thought that the identity relation was not a universal relation, and neither was "x = a" for any object a. Early on, Armstrong wrote that

> Every particular is identical with itself. So the predicate 'is identical with itself' applies to each particular. But we are not thereby forced to admit that particulars have a property, *being identical with themselves.*

and he continued:

> Two reasons may be given for denying that there is any such property.

which makes it evident that that refusal to introduce identity relations runs deep. His thought about the identity relation is forcefully stated in his discussion of the formal properties of the relation N(F,G) that he attributes to laws

> (1) First, the relation is necessarily irreflexive. a's being F cannot necessitate a's being F, nor can there be a law of nature of the form N(F,F). This seems fairly obvious in itself. But is any case I believe that nothing is genuinely related to itself (Armstrong, 1978, Ch. 19, Sec. VI).[10]

Thus Armstrong's account of laws, with its heavy reliance on his account of universals, turns out to be resistant to friendly amendments that were designed to make up for its expressive inadequacy. The addition of (**C**) together with (**MP**∗) would allow a justification of his view that laws are logically stronger than universally quantified material conditionals. However, the notion of predication that (**C**) introduces is too weak to allow an identity relation to be defined. And we see that that the identity relation cannot even be added to Armstrong's account because of his refusal to count the identity relation as a universal. That leaves his account of laws in an undesirable state where it cannot express some of the most significant laws that are on record. It's not a happy state of things.

---

[10]Armstrong, *What is a Law of Nature*, Cambridge University Press, 1983.

## 6.3 Examples of Second-Order Laws and Their First-Order Instantiations

There is the problem of seeing whether there are some simple examples that illustrate Armstrong's claim that laws have the form N(F,G) of a second-order universal. Since laws are, according to Armsrtrong, second order universals, the implication is that they will have instantiations that are first order universals.

There are two examples that Armstrong offered in a later work.[11] *A World of States of Affairs*, Cambridge University Press, 1997. One is Newton's Law of Gravitation, and the other concerns his Guillotine and Decapitation example. They are worth quoting at length, though, I shall replace his Gravitation example in favor of Galileo's Law of Falling bodies only because it is a bit easier to follow, as illustrations go. The Gravitation example is this:

(I) "The gravitational force holding between bodies depends upon the product of the two masses divided by the square of the distance. This gives us three different quantities (two of them masses) that can vary independently. For each different triple of numbers that measure these quantities one can calculate a force. One can think of this calculation as yielding a gravitation law that holds for just these two mass-values and just this distance value. There are innumerable, or perhaps infinite, numbers of these determinative laws. They involve determinate masses and distances, which we may assume are genuine universals, strictly identical in their different instances. (For simplicity of discussion, we accept the fundamental concepts of Newtonian physics as picking out true universals.) What we appear to have in these determinate laws are first-order laws linking first-order determinate universals. . . .

If however, the functional law is a *unitary entity* of some sort, then all the determinate laws can be deduced from the existence of that law plus the assumption of appropriate antecedent conditions – viz. two objects with certain masses at a certain distance. And not only is there deduction, but there is *explanation* because of the unification that the determinable law brings to the huge class of determinate laws. "It's the same damn thing going on in each case." (WSA, pp. 242–243)

And the Guillotine example is given by

(II) "Now consider the guillotine, a, coming into suitable relation to a person, b, and so causing b's immediate decapitation. This is an instantiation of the law, an instantiation of the structural universal that that is the linking of certain universals.

The suggestion is then this. There is *nothing* to the law except what is instantiated in such sequences. Each sequence is an instantiation of that strictly identical

[11] *A World of States of Affairs*, Cambridge University Press, 1997.

'thing' or entity: *guillotining causing the guillotined to be immediately decapitated.* The law, a certain sort of universal, has no existence except in the particular special sequences." (WSA, 227).

In each case, Armstrong considers the determinate laws, that result when, in the case of the law of gravitation, one replaces the three universals by specific values for the masses of each of the two bodies and the distance between them, and then, calculates the gravitational force using the triple of those three specific numbers. The result is supposed to be an instantiation of the second order universal that he takes to be Newton's law of Gravitation. There are, he says an infinite number of these determinate laws obtained by these instantiations, and in each case one obtains, he says, a first-order relation between universals that is also a first order law between universals. So the claim is that there are many first-order laws, each of which is an instance of the Newtonian Law of Gravitation, and also many beheadings upon the actions of a guillotine, each of them being a law.

It is far from obvious why he described the instantiations of the second order universals (laws) as laws ("determinate laws"). I think what he has in mind can be unpacked with the help of two of his examples.

Consider Armstrong's guillotine example first: there's a guillotine a, and a coming into a suitable connection (sic) with a person b (presumably one state of affairs) standing in what seems to be a causal relation to a second state of affairs, b's capitation. So we have two states of affairs standing in a causal relation, as being an instance of the law. What law is that supposed to be? Armstrong clearly thinks that there are many laws of succession that involve the actions of a guillotine.[12] The succession in each case is clear, but the generalization of what takes place in each of the various instantiations is not a law, nor is the succession, in the case of any particular beheading, a law. The case of Lavoisier losing his head is not a first- order law, nor is it a deductive consequence of the generalization about the action of guillotines and unshielded necks, as Armstrong thinks. It's just one specific unfortunate sequence after another. This example is not a happy one.

However the example of the Newtonian Law of Gravitation, or the simpler one of Galileo's Law of falling bodies doesn't fare any better. Either case is an example of what Armstrong calls a functional law. The Galilean law states that in a vacuum, a body falling freely (in a vacuum) towards the Earth such that if x is the vertical distance fallen in an elapsed time t, then

$$\text{(G)} \quad x = (1/2)gt^2,$$

where g is the gravitational constant. This is supposedly represented somehow by the relation N(F, G) between two universals. The instantiations of this second- order universal are supposed to be obtained by substituting a specific distance xo and a specific time to, in the expression of the law. The result will be one among the many,

[12]For more detailed discussion of this example Cf. WSA, 225 ff.

posssibly infinite instantiations of the Galilean Law. However, each of his instantiations is not a law, but a simple equation with specific values for "x", and "t". i.e.

$$(\text{Go})\ \ \text{xo} = (1/2)\ \text{gto}^2.$$

(Go) tells you only how in one particular case, the distance fallen in an elapsed time is related to a specific time that has elapsed. That is a particular fact. It is not a law.

Armstrong claimed that the results of these instantiations are laws, and with special assumptions about the values in the functional laws, one can *deduce* determinate laws. That is, given (G) together with assumptions about the special values, one can deduce (Go). Furthermore, this deduction he believes *explains* the determinate laws. He is surely right about the conditions under which (G) will yield (Go), but is wrong that the instantiations are laws. (Go) is not a law. The conclusion we drew in the guillotine case parallels our conclusion for the Galilean case: we wind up with a possibly causal relation between two *specific* states of affairs, related by some binary universal – say R.

## 6.4 Some Formal Properties of the Relation of Necessitation N(F,G)

Armstrong described some formal properties of the N(F, G)s without anything like a proof that they had those properties. They are second-order universals because they are relations between first-order universals. Consequently, without additional information about these N(F, G)s, it is not at all obvious how Armstrong's claims about their formal properties can be established.

We can therefore try to rectify this gap with the addition of a condition, (i)- below, that will enable us to provide some proofs for the formal conditions he discussed.[13]

In setting forth the additional conditions we will rely heavily upon what he said in paragraphs (I) and (II) above, which concerned the connection of N(F,G)s with those sequences of states of affairs associated with the universals F and G. Actually there are many ways of expressing the connection. Here is one possibility which reflects ideas expressed in (I) and (II):

(i) $\text{N}(\text{F}, \text{G}) \Leftrightarrow (\text{x})\text{R}(\text{F}[\text{x}], \text{G}[\text{x}])$.[14]

---

[13]It should be noted that these four formal properties endorsed by Armstrong cover only the case when the laws involve only two universals.

[14]We have to be careful in attributing this equivalence to Armstrong. He does say that laws are second order universals relating (first-order) universals, and he also says that laws relate states of affairs by a binary second order universal. So (i) should be seen as a friendly addition that is not explicitly advocated by Armstrong. We do not claim that (i) is the only was of glossing the connections between Armstrong's account of laws and states of affairs. There are many variations on this theme consistent with what one can glean from (I) and (II).

Where the double-shafted arrow indicates logical equivalence. (i) says that N(F, G) is logically equivalent to the statement that for every particular x, the state of affairs, F[x], (x's being F), is related by the binary universal R to the state of affairs G[x] (x's being G),[15]

It is this condition that seems to gloss Armstrong's claim in (II), that

> The law, a certain sort of universal, has no existence, except in the particular special sequences.

From (i) it also follows, that for any N(F, G),

(5) $N(F, G) \Leftrightarrow (x)\ R(F[x], G[x])$, and hence

$$(6) \quad \begin{aligned} N(F,G) \Rightarrow\ & R(F[a]), G[a]), \\ & R(F[b], G[b]), \\ & R(F[c], G[c], \\ & \ldots \end{aligned}$$

Armstrong thought that each of the cases listed in the right-hand side of (6) are the determinate cases of the determinable law on the left-hand side. (6) would also provide a way of understanding Armstrong's assertion (in (I)), that

> ...because of the unification that the determinable law brings to the huge class of determinate laws. It turns out to be the same damned thing going on in each case.

We are now in a position to provide arguments for several formal conditions that Armstrong thought that the N(F, G)s, should satisfy.

We shall assume that the binary universal R used in (i) is irreflexive, not symmetric and not transitive on the set W of all states of affairs. We then have the following results:

Irreflexivity: It is never the case that N(F, F) for any universal F. Suppose that N(F,F). Then by (i), we have N(F, F) $\Leftrightarrow$ (x)[R(F[x], F[x]). Since R is irreflexive on W, we have the negation of R(F[x], F[x]). and consequently (by (i), the negation of N(F,F).

Non-symmetry: It is never the case that N(F, G) $\leftrightarrow$ N(G, F) for every F and G. Suppose that N(F, G)) $\leftrightarrow$ N(G, F)), for every F and G. By (i) then N(F, G) $\Leftrightarrow$ (x)R(F[x], G[x]), and N(G, F) $\Leftrightarrow$ (x)R(G[x], F[x]). Therefore (x)R(F[x], G[x]) $\leftrightarrow$ (x)R(G[x], F[x]). Since R is not symmetric on W, it follows that N(F, G) is not symmetric.

---

[15]R is a binary universal that Armstrong thinks of as some kind of causal relation. It should be noted that here, unlike familiar notions of that relation, it is not a relation between objects, events or facts, but a relation between states of affairs. It should also be noted that some, writers, eg H.G. Mellor do not regard causation as a relation and R.B. Braithaite (in his Fellowship Thesis for King's College, *A Dissertation on Causality* (1923), King's College Archives, Cambridge, was one of the first to use the causal relation in hypotheticals (asserted conditionals) merely to indicate some kind of relation where one item depends upon another, leaving it open that it could cover various kinds of determination. Armstrong owes an explanation of the relation R between states of affairs, that figures so prominently in his account of laws.

Non-contrapositive. This condition, that N(F, G) $\leftrightarrow$ N(not G, not F), always fails for the simple reason that there aren't any negative universals.[16]

Non-disjunctive. There are no disjunctive laws -i.e. that is: no cases of N(U, V), where neither U nor V is a disjunction of universals. That is because there are no states of affairs that are disjunctive.[17]

Non-transitive. That is, N(F, G) and N(G, H) do not imply N(G, H), for all universals F, G, and H.

Suppose that N(F, G) and N(G, H). Assume, in addition that N(F, H). Then, by (i) we have,

$$(x)R(F[x], G[x]), (x)R(G[x], H[x]), \quad \text{and} \quad (x)R(F[x], H[x]).$$

Consequently, we also have

$$R(F[a], G[a]), R(G[a], H[a]), \text{and} \quad R(F[a], H[a]).$$

This conclusion however says that the state of affairs H[a] is causally dependent on the state of affairs F[a} and is also causally dependent on the state of affairs G[a]. This is of course, a case of causal over determination. Assuming that causal determination is false, non-transitivity condition always fails.[18]

We think that our additions to Armstrong's account help to advance it a bit by providing some support for some of the deficiencies that have been noted in the literature. However, his account still falls seriously short of being satisfactory. Not only does it fail to certify some central laws as such; it cannot express any law that requires magnitudes for its expression. The reason in short, is that laws involve physical magnitudes, and those, since they are functions, they require an identity relation for their expression.

In addition things seem even worse. We were able to prove that the N(F, G) satisfy the formal conditions of irreflexivity, non-symmetry,, non-contrapositivity, non-disjunctivity, and non-transitivity. This involved the aid of the exrtra assumption (i). The proofs are proofs of formal conditions of the N(F,G)s which Armstrong endorsed. That is a plus for his view, but that view is, we think, mistaken.

With the exception of the condition of irreflexivity, the remaining formal conditions are mistaken. Briefly: (1) Contrapositivity, as we noted, holds for those laws which are conditonal since they are equivalent to their contrapositives; non-symmetry fails. (2) Symmetry holds for those laws that express an equation between physical magnitudes, say $M = M*$. Then M if and only if M$*$. So symmetry prevails rather than nonsymmetry in such cases. (3) The claim that there are no disjunctive laws is questionable. Consider Newton's first law phrased this way: If

---

[16] *A World of States of Affairs*, Cambridge University Press, 1997, p. 19.

[17] Ibid.

[18] Armstrong seems to have held that ontological causation is non-transitive, though its transitive closure is transitive. Cf. *A World of States of Affairs*, 206–207.

there are no forces acting on a body, that that body is in one of two states – either it is in a state of rest R with respect to an inertial frame I, or it is in a state of uniform motion U with constant non zero velocity with respect to I – if we think of "being in one state of motion or being in another state of motion", as a disjunction of universals or a disjunction of states of affairs. We could describe the situation so that it looks like this: If a body has no forces on it then it instantiates the universal of being at rest (relative to an inertial frame I, or it instantiates the universal U of having a constant non-zero velocity in I. If each of these disjuncts counts as a state of affairs, or, if each of them is a universal, then that is a problem for Armstrong since, on his account of them, there are no disjunctions of universals, and no disjunctions of states of affairs.[19] (4) The proof that we gave for non-transitivity comes down to the claim that there is no overdetermination of states of affairs.

We have thus far concentrated on some familiar and influential accounts of laws and found them to be inadequate despite their ingenuity. We now wish to make a fresh start by presenting a mini-theory that incorporates a concept of explanation, and consequently departs from the older received view.

[19] *A World of States of Affairs*, p. 19.

# Chapter 7
# Laws and Accidental Generalities

From the beginning of philosophical interest in laws and explanation, the emphasis was on laws as playing a fundamental role in explanations. This was evident in Aristotle (if one understands that his reference to four kinds of causes should be understood as his interest in four kinds of explanations.) In our time, the emphasis was very clear in C. Hempel and P. Oppenheim's seminal essay (Cf. Chap. 1).

In the present chapter we will depart from that tradition and suggest that explanations have a fundamental, role in framing the important, but neglected distinction between laws and accidental generalizations. The dividends of such a departure is, we think, worth the trouble. Depending upon the features that are ascribed to explanation one can show, among other things, that (1) No laws are accidental, (2) All laws are true, and strikingly, (3) Any explained contingent generalization is not accidental.

## 7.1 The Lotze Uniformity Condition

It has become a hallmark of any satisfactory account of scientific laws, that they cannot be accidental generalizations. There is of course a very easy way of assuring the difference, and that is by defining accidental generalizations as exactly those

---

It is a great pleasure to dedicate this chapter to Pat Suppes. It it is based upon "Laws, Accidental Generalities, and the Lotze Uniformity Condition", in *Conceptual Clarifications, Tributes to Patrick Suppes (1922–2014),*EDS. Jean-Yves Beziau, Decio Krause, Jonas R. Beccker Arenhardt, Individual authors and College Publications, D. Gabbay, 2015, pp. 175–186. I first came to know Pat's work when I was a graduate student and came across a copy of his dissertation *The problem of action at a distance* (supervised by Ernest Nagel). It was a revelation. The combination of historical accuracy, formal precision, elegance and intrinsic interest had a profound influence on me. When I had a year fellowship to study philosophy anywhere I wanted, I made a bee-line directly to Stanford, then Harvard, and Cambridge. My early work on measurement, and later work in logic bear his imprint, if not his imprimatur. Thanks Pat for showing me the way.



A. Koslow, *Laws and Explanations: Theories and Modal Possibilities*, Synthese Library 410, https://doi.org/10.1007/978-3-030-18846-7_7

generalizations that are not lawlike. That option is something of a cheat. Philosophers who describe a generalization as accidental usually do so to explain why a true generalization is not a law. Thus, Hempel said that the generalization "All the rocks in that box contain iron" failed to be a law because it was accidental that those rocks all share that property. Obviously Hempel didn't just remind us of a definition; he gave a reason.[1] The follow up question then is this: why is the rocks in the box example an accidental generalization? Something more is needed.

On this point J. Carroll has astutely noted that most of the current discussions have focused on scientific laws and not much can be found on accidental generalizations.[2]

Our proposal for marking this distinction among generalizations, has some interesting sources. One immediate source is W.E. Johnson who distinguished two different kinds of generalized conditional: universals of fact, and universals of law:

> It will have been observed that the correlative notions of determination and dependence enter into the formulation of the principles as directly applicable to the characters of manifestations and therefore only derivatively to the manifestations themselves. Hence the potential range for which these principles hold extends beyond the actually existent into the domain of the possibly existent. In this way the universality of law is wider than that of fact. While the universals of fact are implied by universals of law, the statement of the latter has intrinsic significance not involved in that of the former.[3] (*Logic, Part I*, 1921, pp. 251–252).

This is a remarkable observation of Johnson. He says that there are two kinds of uniformity (something which Mill, he notes, failed to separate). He says further, that the universals of fact are applicable to objects (manifestations), and the universals of law apply directly to the properties (characters) of those objects. Moreover, he also required that universal conditionals follow from laws. Finally, he required that all the universals of law should be stated as what we now call counterfactual conditionals.

> Thus taking two determinate adjectives $p$ and $q$ under the respective determinables $P$ and $Q$, the factual universal may be expressed in the form 'Every substantive $PQ$ in the universe of reality is $q$ if $p$; while the assertion of law assumes the form 'Any substantive $PQ$ in the universe of reality would be $q$ if it were $p$.' These formulae represent fairly, I think, the distinction which Mill had in mind; for my first formula may be said to express a mere invariability in the association of $q$ with $p$. while the second expresses the unconditional connection between $q$ and $p$. Or as I have said in p. 252, Chapter XIV, Part I, the universal of fact covers only the actual, whereas the universal of law extends beyond the actual into the range of the possible.[4]

Although Johnson believed that Mill had such a distinction in mind, we think that another more likely source for the contrast between two kinds of conditionals, can be

[1]Prentice Hall, 1966, p. 55.

[2]"Nailed to Hume's Cross" in *Contemporary Debates in Metaphysics*, eds. J. Hawthorne and D. Zimmerman, Blackwell, 2008.

[3]*Logic Part I*, Cambridge University Press 1921, pp. 251–252. Reprinted by Dover Publications, N.Y. 1964a.

[4]*Logic Part III, The Logical Foundations of Science*, Cambridge University Press, 1924. P.6. Reprinted by Dover Publications NY, 1964b.

found in the logical writings of the nineteenth century German philosopher Hermann Lotze, with whose work on logic Johnson was acquainted.

It is not immediately evident from Johnson's writings that he had read Lotze. He is nowhere cited by Johnson. Nevertheless the evidence for this is clear since Johnson was lecturing on Lotze's *Logik* in 1888, in his Cambridge course on logic. At that time, in Cambridge, lecturers listed the syllabus for their course of lectures on a printed formal card, a brochure of a sort, which was presumably made available to students. There is in the Cambridge University Archives, a volume of the minutes of the Board of Research studies, 1888, of a copy Johnson's printed card announcing the books that he had chosen as texts for his course on logic which included Lotze's *Logic*.[5]

Lotze anticipated the distinction between two types of conditionals which he called *general* and *universal,* and that we find reflected in Johnson's *Logic*. It is to Johnson's credit that he thought that laws should be represented as conditionals of the second kind (Lotze's universal conditionals), while Lotze thought that there was a connection between his universal conditionals and their non-accidentality.

We turn then to a more detailed consideration of Lotze's observations, and to a way of sharpening his insight into two principles that connect up lawfulness and accidentality, with the help of a condition which we shall call the *Lotze Uniformity Condition* (LUC).

Here is the seminal passage from Lotze's *Outlines of Logic and of Encyclopedia of Philosophy*, which raises the consideration of "unfortunate accidents".

> §32. This thought gains expression in the form of the *general* judgment. Such form is to be distinguished from that of the *universal* judgment. The latter of the form only asserts that, in fact, all instances of S have P, -for example, 'All men are mortal,' – but does not tell why. Perhaps it may be on account of a combination of unfortunate accidents which have no real connection with each other.
>
> All S are P
>
> The general judgment substitutes the general concept alone for the subject: 'Man is mortal'; or it indicates by the other form, 'Every man is mortal,' that the predicate is to be considered valid, not merely of all actual but also of all thinkable examples of S; and therefore is so by virtue of this same general concept S, and not on other accidental grounds.
>
> More accurately considered, the general judgment must besides be included in the hypothetical form. For it is not the general concept S (the universal man) which is to be considered as P (mortal); but every individual, *because* he is a man. Therefore, the general form, strictly speaking, is; 'if any A whatever is an example of the universal S, then such A is necessarily P.[6]

A few remarks may make his point stand out more clearly. Lotze distinguished between *general* judgments and *universal* ones. The universal ones seem to be just

[5]My deep thanks to Elizabeth Leedham-Green, lately Cambridge University Deputy Archivist for finding the brochure, with its clear link to Lotze.

[6]*Outlines of Logic and of Encyclopedia of Philosophy*, ed. and trans. G. T. Ladd, Boston, MA: Ginn & Co., 1887.

the sort of generalized conditionals that can be represented using a standard first-order universal quantifier: $(\forall x)(Sx \rightarrow Px)$. The second kind, the general conditional, differs from the factual one in two ways. The first concerns the scope of the universal quantifier. Lotze thought that in the universal judgment the quantifier concerned all those things which are *actual* Ss, while in the general judgment the quantifier concerns not only all actual things which are Ss, but all *thinkable* things which are Ss.

When F. Ramsey noted this distinction as it was expressed in Johnson's logic, he spotted the ambiguity. Is the difference one of two kinds of quantifier, or one of two kinds of conditional. It looks as if Lotze and Johnson both framed the difference in terms of different universal quantifiers. And that prompted Ramsey's remark that Johnson lacked an understanding of the universal quantifier. Johnson, he said, didn't understand that "everything" means everything. However, the difference between the two kinds of judgment gets to the heart of Lotze's concept of accidental judgments. The general judgments are not accidental.

Lotze says that the predicate (mortal) holds of all and possible things in a non-accidental way that hints of explanation: the judgment he said, requires that every instance of a man be mortal *because* of the uniform reason of being a man. His insight was that it was not enough for a law to be a true universal generalization: "All Fs are Gs." Something in addition to truth is needed, since it might just be an accident that the generalization is true. For suppose that indeed, everything that is an F is a G. That could have come about this way: each particular F is a G, but the reason that each F is G might be different for different Fs. If a generalization is accidentally true, then what happens is that the generalization is true, but some of its instances hold for different reasons.

What this suggests is that the truth of the instances of a law are all insured by some uniform factor. This requirement could be understood in at least three ways: explanatory, causal, or rational, – that is the provision of a single explanation, a single type of cause, or a single type of reason for all of the instances. Here we shall consider only the explanatory option.

A few terms are needed to express what I shall call *Lotze's Uniformity Condition* (**LUC**). We will use the expression "Exp[A; B]" to say that *there is* some explanation of B, that is provided by A. This is an existential claim and not a reference to any particular explanation that makes" Exp[A; B]" true.[7] The particular explanation might be provided by an argument, a cause, or just some true statement. For our purposes, it doesn't matter what particular model or account of explanation one might advocate. We can now formulate the Lotze Uniformity Condition (**LUC**) this way: If S is the generalization (say) All Fs are G, then **LUC**(S) $\Leftrightarrow$ For *some* R $(\forall x)$ Exp[R; Fx $\rightarrow$ Gx],.[8]

[7]Although we do not assume that all explanations are deductive, our use of the relation "Exp "is similar to the use in logic of the relation of deducibility "$\vdash$", in that an existential statement is intended. "A, B $\vdash$ C" means that *there is* a deduction of C from premises A, and B. It is existential and does not refer to any particular deduction that does the job.

[8]Here, and in what follows, we use "$\Rightarrow$"to indicate logical implication, "$\Leftrightarrow$"for logical equivalence, and "$\rightarrow$"for the material conditional.

The phrase "*for some* R" is intended to have wide scope. Thus the universal conditional S satisfies the Lotze Uniformity Condition if and only if there is some R that provides an explanation of all the *instances* of S.[9]

With this in place we can now construct a mini-theory of two conditions that will connect up the notions of laws and accidental generalizations with the Lotze condition. We say this to emphasize the fact that at this stage, we are considering something short of a definition.

For any generalization S (eg. All Fs are G), and using "$\mathcal{L}$" to indicate the predicate "it is a law that . . .", let the conditions (1) and (2) constitute the mini theory of Laws and Accidental generalizations (**LAG**):

$(1*)\ \mathcal{L}\ (\mathrm{S}) \Rightarrow \mathrm{LUC(S)}\ \big(\mathrm{For}\ \textit{some}\ \mathrm{R}\big)\ (\forall \mathrm{x})\ \ \mathrm{Exp[R;\ \ Fx \rightarrow Gx]}\big),$ and
$(2*)\ \mathrm{Acc(S)} \Rightarrow\ \neg\ \mathrm{LUC(S)\ (For\ no\ R)\ (\forall x)\ \ Exp[R;\ \ Fx \rightarrow Gx])}$

There are a few observations and consequences worth noting. The first condition requires that Lotze uniformity is a necessary condition for being a law: If S is a universal conditional, then some R explains every instance of it. The second condition is a necessary condition for a universal conditional to be an accidental generalization: "S is an accidental universal conditional" implies that there is no R which explains every instance of it.

Although the Lotze conditions were originally proposed for generalizations of the type All Fs are G, the uniformity condition can easily be extended to cover all generalizations of the form $(\forall x)\Omega(x)$, just as long as it makes sense to speak of the instances of $\Omega$. Consequently, it is worth emphasizing that the laws and accidental generalizations covered by (1) and (2) can be extended to a much broader class than just universally quantified conditionals.

Here are some of the more elementary consequences of (LAG) including those which depend on features of the concept of explanation that is used in the expression of the Lotze uniformity condition.

## 7.2 Observations and Consequences for (LAG)

(i) *Laws and accidental generalizations. No generalization S can be both a law and an accidental generalization.*

[9]We have used the conditional [Fx → Gx] to indicate the form of the instances of a law that is conditional in form. This differs from the early discussions of confirmation in which instances were assumed to have the form Fx &Gx. That led to the unacceptable consequence that logically equivalent formulations of a law would not have the same instances. Here I follow Hempel's later use of "instance" that blocks that consequence (*Aspects of Scientific Explanation*, the Free Press, 1965, p. 341, footnote 7.

This is an immediate consequence of (**LAG**) that most writers regard as central to any account of scientific laws: That is, all laws are not accidental:

$$\text{For every generalization S}, \mathcal{L}\ (\mathrm{S}) \Rightarrow \neg \mathrm{Acc}(\mathrm{S}).$$

It is a little surprising that this result can be obtained without any deeper analysis of either the notion of scientific law or explanation. If we are correct, then this account explains why no generality can be both a law and accidental. Moreover, this account can certainly be a part of any Humean view of laws. There is a caveat, however, if Humean accounts of laws are supposed to eschew any dependence on modal notions. In that case there might be an objection on two grounds: that explanation is a modal concept (I believe that it is) and that the statement that S is a law ($\mathcal{L}$(S)), is itself a modal statement. These observations even if correct, do not impugn a Humean account of laws. One can surely write about explanations and laws without impugning anyone's Humean credentials. Otherwise a Humean account, even if one wanted to give one, would be impossible. We think however, that some modals are part of any Humean's legitimate philosophical lexicon.

(ii) *Laws and their instances*. Condition (1) should not be confused with a requirement that is sometimes defended in the literature – that it is only laws that are confirmed by their instances. That is a view usually associated with Nelson Goodman. When coupled with the idea that the explanation relation is just the converse of the confirmation relation, one could conclude that laws explain their instances. This is a result that would make (**LAG**) (1) obvious. The reason is just that if S is a law then there is always some R which explains all of its instances – namely S itself. This claim, even if true, is irrelevant to Condition (1), because the notion of "instance" that Goodman used is different from the notion of "instance" that is used in (**LAG**), as was already noted.[10] Consequently, Condition (**LAG**) (1) gets no support from the old claim that laws explain their instances. What (**LAG**) (1) says is that if S is a law, then there is *some* R that explains all its instances (in our sense of "instance"). It does not require that the uniform explanation be given by the law S.

(iii) *The converses of the conditions of (LAG)*. We have not assumed the converses of (**LAG**) (1) and (**LAG**) (2), though both raise interesting possibilities worth exploring. According to the converse of (1), if there is a uniform explanation of each of the instances of a regularity, then that implies that the regularity is a law. The result would be that for any generalization S, S's being a law ($\mathcal{L}$(S)) and S's satisfying the Lotze condition (**LUC**(S)) would mutually imply each other. This looks initially attractive since it provides something like a definition of the predicate "is a law". However we shall note below in our discussion of Reichenbach's example of golden cubes, that in combination with some assumptions about explanation, it has some moot consequences. We shall

[10] Chapter 2, footnote 7.

consder this when we turn to those consequences of **(LAG)** (1) and **(LAG)** (2) when they are combined with some additional assumptions about explanation.

Thus far we have considered consequences of the theory **(LAG)** that do not rely upon any properties of the notion of explanation used in the expression of the Lotze Uniformity Condition. We turn next to a number of consequences that are obtained with the help of some general assumptions about explanation.

(iv) *All laws are true*. That is, $\mathcal{L}(S) \Rightarrow S$. This follows from (LAG) together with the assumption that explanations are factive. By "factivity" we mean that for any A and B,

$$\text{Exp}[A; B] \Rightarrow A, \text{and } \text{Exp}[A; B] \Rightarrow B.$$

I.e. "There is an explanation that A provides for B" implies A as well as B. This factivity condition is two-sided. The usual examples, such as "Richard knows that P" implies P are so-called one sided factives. In the case of explanation, both propositions are implied. In short, there is no explanation of B that is provided by A, unless both A and B are true. Here then is the straightforward proof that the prefix "It is a law that ..." is factive."

Let S be the generalization $(\forall x)\Omega x$. Suppose now that S is a law – i.e.$\mathcal{L}(S)$. By **(LAG)** (1), we have **LUC**(S). Therefore $(\forall x)\text{Exp}[R; \Omega x]$, for some R. By factivity, $\text{Exp}[R; \Omega x] \Rightarrow \Omega x$. Therefore $(\forall x)\text{Exp}[R; \Omega x] \Rightarrow (\forall x)\ \Omega x$, for some R. The antecedent is just **LUC**(S), and the consequent is just S. Therefore $\mathcal{L}(S) \Rightarrow S$. Thus the result is this: the sentential operator "It is a law that ..." is factive.

(v) $\mathcal{L}(S)$, *"It is a law that (S)" is non-extensional*. Here we shall sketch a proof for the case where "S" is conditional – say for example, $(\forall x)(Fx \rightarrow Gx)$. If it is a law that $(\forall x)(Fx \rightarrow Gx)$, then we have that $\mathcal{L}((\forall x)(Fx \rightarrow Gx))$ implies that $(\forall x)\text{Exp}[R, Fx \rightarrow Gx]$ for some R. If we assume that "Exp" is nonextensional, then let $F{*}x$ be a predicate that is coextensional with Fx, such that $\text{Exp}[R, Fx \rightarrow Gx]$, but not $\text{Exp}[R, F{*}x \rightarrow Gx]$. Now we know that if it is a law that $(\forall x)(F{*}x \rightarrow Gx)$, then we have that $\mathcal{L}((\forall x)(F{*}x \rightarrow Gx))$ implies that $(\forall x)\text{Exp}[R, F{*}x \rightarrow Gx]$ for some R. But by hypothesis, the consequent is false. Therefore it is false that $\mathcal{L}((\forall x)(F{*}x \rightarrow Gx))$. Consequently, "$(\forall x)(F{*}x \rightarrow Gx)$" is not a law. This was already known from our discussion of Dretske's example of mining in Kimberlite (Chap. 4), and reinforces our argument given there, that it is not the law that all diamonds have a certain refractive index that is nonextensional. It is the claim that that the prefix "it is a law that " is nonextensional. The particular example is telling, but it is nice to have an argument for the general case (assuming of course that the nonextensionality of explanations is assumed).

(vi) *Explained generalizations*. The following result answers a simple question: What happens if a generalization is explained? *If a generalization S is explained, then S is not accidental.*

The answer is surprising, but, I think, welcome. If one assumes that "Exp" is closed under implication, then, if there is some explanation of a generalization, then that generalization satisfies the Lotze uniformity condition, and consequently it is not an accidental generalization.

The assumption of closure for explanation is admittedly controversial, though I have defended that requirement elsewhere.[11] We shall assume that Exp is *closed under implication*. That is, for any R, A, and B,

$$\text{If } A \Rightarrow B, \text{then } \mathrm{Exp}[R; A] \Rightarrow \mathrm{Exp}[R; B].$$

Briefly, if A implies B, then "There is an explanation that R provides for A" implies that there is an explanation that R provides for B.[12] The reason is fairly direct. Suppose that S is a generalization, say $(\forall x)\Omega(x)$. Since $(\forall x)\Omega(x) \Rightarrow \Omega(a)$ for any instance a of $\Omega(x)$, we have, by closure of Exp, that $\mathrm{Exp}[R; (\forall x)\Omega(x)] \Rightarrow \mathrm{Exp}[R; \Omega(a)]$ for all for all instances a. That is, $\mathrm{Exp}[R; S] \Rightarrow \mathrm{LUC}(S)$. However by (2) of (LAG), LUC(S) implies $\neg$Acc (S). Consequently, $\mathrm{Exp}[R; S\} \Rightarrow \neg\mathrm{Acc}\ (S)$. That is, if any generalization has an explanation, then it is not accidental.

Lest the assumption of the closure of explanation be thought too controversial, we note that the same conclusion can be obtained with the use of the much weaker assumption of *restrictive closure*. We shall say that an explanation relation satisfies restrictive closure if and only if:

(**RC**) If S is a contingent generalization and $S*$ is any instance of S, then for any R, $\mathrm{Exp}[R; S] \Rightarrow \mathrm{Exp}[R; S*]$.

The argument now runs this way: Let $(\forall x)\Omega(x)$ be a contingent generalization and suppose that $\mathrm{Exp}[R; (\forall x)\Omega(x)]$. Let $\Omega*$ be any instance of $(\forall x)\Omega(x)$. By restrictive closure, $\mathrm{Exp}[R; \Omega*)]$. So there is a uniform explanation (namely R) of all the instances of the generalization. Therefore $(\forall x)\Omega(x)$ satisfies the condition (LUC). Consequently, $(\forall x)\Omega(x)$ is not accidental.

There is an interesting moral that can be drawn from this result. It is a widespread though, not universal belief, that explanations are important and are sought, and prized. Why is this so, assuming that it is? One answer provided by the preceding result is this: It is one thing to have a generalization that is true. Roughly speaking, that tells us the way the world works. If the generalization is also a law, then by (LAG), that tells us even more about the way the world works. However, even if a true generalization S is not a law, nevertheless, an explanation of S tells us

[11]"Explanation and Modality", A. Koslow, typescript, and as a talk in Durham.

[12]This form of closure is weaker than it may appear. It does not require that if A implies B, then there is an explanation of A that is also an explanation of B. The explanations maybe different. Indeed the deducibility relation is a case where for different A, and B, such that A implies B, the closure condition holds. Nevertheless any specific proof (deduction) of A from some C will not be a proof (deduction) of B from C. We are not suggesting that explanations are proofs; only that they both satisfy the closure condition.

that S satisfies the Lotze uniformity condition, and that consequently, it is not accidental. That tells us more than that it is just a true generalization.

## 7.3 Laws as Bound with Explanation

According to the older, but mainstream tradition, explanations (whether deductive or probabilistic) are dependent upon some account of law. Accordingly, any explanation must involve an essential use of one or more laws; probabilistic laws for probabilistic explanations, and non-probabilistic laws when the explanation is non-probabilistic. The kind of explanation will determine the kind of law that is needed. The traditional program required that some account of laws should already be in place. Our mini theory (LAG) involves a departure from that program. It requires that some concept of explanation be already in place in any account of laws. I think that the proper conclusion is not that one of the two concepts, that of law and that of explanation is in some sense "prior" to the other. The proper conclusion is that the concepts of laws and accidental generalizations on the one hand, and explanation on the other, are interwoven, and have to be developed in tandem.

In this connection there are two accounts of laws and accidental generalizations that deserve closer study because they also obtain interesting results by exploiting a connection between laws and explanation. One is a recent account by John Carroll, and the other a much earlier account by Richard B. Braithwaite. In "Nailed to Hume's Cross?"[13] Carroll offers an account of laws and coincidences, according to which,

> "P is a coincidence if and only if there is no Q such that P because Q" (73), and "P is a law of nature if and only if P is a regularity caused by nature. (74).

He explains that by "caused by nature" he means "that the law is true *because* of nature"(emphasis by J. Carroll); not that nature is somehow a causal agent. The intention is that "cause" should be understood in the explanatory sense that "cause" can have. This is a very interesting incorporation of explanation into an account of laws. The key notion of a regularity that is caused by nature is supposed to convey that the regularity is true because of *nature itself.* As Carroll says,

> Lawhood requires that nature itself – understood as distinct from anything in nature or the absence of something from nature – make the regularity true. (74)

This would seem to imply that all laws are explained regularities – a very significant claim. Moreover all the laws have at least this one explanation –nature. Those of us who look to theories for explanations of laws, and who also think that there may be

[13]Contemporary Debates in Metaphysics, J. Hawthorne, T. Sider, D. Zimmerman (eds.), Oxford, Basil Blackwell, 2008.

different explanations (when possible) for different laws, might be disappointed because that kind of specificity isn't mentioned. It is easily supplied, and ought to be regarded as the fuller account Carroll's intended.[14]

There are some very nice features that Carroll's theory yields. For example, it yields the result that laws cannot be accidental regularities, and it makes it clear that it is explanation that makes the difference between those regularities that are laws, and those that aren't. Clearly this is a theory that deserves further development and refinement.

Richard Braithwaite has also offered an account of laws and accidental generalizations which exploits their relation to explanation.[15] He used the notion of well-established *deductive systems* which are algebraic structures associated with theories, that are neither the set of axioms for the theory nor are they Tarskian theories (sets closed under implication). He defined laws of such systems in such a way that if H is an hypothesis in a deductive system $\mathcal{D}$, and H is explained in a deductive system $\mathcal{D}*$ which properly includes $\mathcal{D}$, then H is a law of the system $\mathcal{D}*$.[16] That is a seemingly stronger result than the one that we obtained: that any explained generalization is non-accidental. However, the results are nearly the same, given that Braithwaite regarded accidental generalizations as those for which there is no established scientific deductive system in which the generalization appears as a consequence (304). Thus he could have equally well concluded that under his conditions, the generalization is non-accidental.

There are however some significant differences that ought to be noted: (1) For Braithwaite, laws are always part of some established deductive system $\mathcal{D}$ that is associated with some theory, whereas the account in (LAG) seems to consider laws independently of some theory or deductive system. Another way of indicating the difference is that for most accounts of scientific laws, the question of whether P is a law of $\mathcal{D}$ is determined by two clauses: (1) is P a law, and (2) is P a member of $\mathcal{D}$. Braithwaite's idea is that whether P is a law or not is concerned only with whether (2) is true for some deductive system $\mathcal{D}$. In focusing on the problem of whether a law is a member of some deductive system, Braithwaite may be right.[17] That insight however can be incorporated in a way into (**LAG**) by adjusting the Lotze uniformity condition. Instead of the requirement that some R explains all the *instances* of a generalization S, we now require the more refined condition that there is some *theory*

---

[14]This is clear from private correspondence with Carroll.

[15]*Scientific Explanation*, Cambridge University Press, 1953.

[16]An account of Braithwaite's notion of a deductive system, laws of deductive systems, and explanations of laws in a deductive system involves several conditions of the definite inclusion of one deductive system within another. A more detailed discussion of Braithwaite's account is developed in Chapter 8, (IV).

[17]A good example of this kind of account of laws that stresses the deductive role that a law plays within a scientific theory, but is otherwise very different from Braithwaite's is the richly detailed recent one of John T. Roberts in *The Law-Governed Universe*, Oxford University Press, 2008. Both are akin to, but different from our focus on laws that have a theoretical context, which we call the *theoretical scenario* for the laws, and which is discussed in the chapters which follow.

T that explains all the *instances* of a generalization S. Call this the theoretical version the Lotze Uniform Theory condition: (**LUCT**). That is, for any generalization S say for example, All F's are G,

$$\mathbf{LUCT}(S) \Leftrightarrow \text{For } \textit{some} \text{ theory T, } (\forall x)\ \text{Exp}[T; Fx \rightarrow Gx],.$$

This seems to us to be a natural refinement of the version that only asked for *some* explanation. After all we are presently interested in the explanations of contingent generalizations and their instances, and that is rightly thought to be the work of theories.

The idea then is to use the theoretical version of the uniformity condition of our mini-theory (**LAG**). Replace "(**LUC**)" by "(**LUCT**)", so that our mini-theory is now given by (**LAGT**):

(1∗) $\mathcal{L}$ (S) ⇒ **LUCT**(S) (For *some* theory T, (∀x) Exp[T; Fx → Gx]), and
(2∗) Acc(S) ⇒ ¬ **LUCT**(S) (For no theory T, (∀x) Exp[T; Fx → Gx])

We turn now to an application of this theory-bound version of the uniformity condition to see how well it squares with some familiar examples of laws and non-laws in the literature.

## 7.4 Gold Spheres, Rusty Screws, and Rocks with Iron

We will consider those examples, first with the use of the uniformity condition that requires that there be some R that explains each of the instances of a law (LAG), and then a more refined version of the uniformity condition that requires that there is a *theory* T that explains all the instances of the generalization, i.e. (**LUCT**).

Consider three classic examples of regularities that fail to be laws: H. Reichenbach's

(**G**) "All gold cubes are less than one cubic mile in volume.", E. Nagel's
(**N**)"All the screws in Smith's care are rusty",[18] and C. Hempel's
(**H**) "All the rocks in this box contain iron.".[19]

The uniform conclusion that none of these examples are laws is usually supported by appealing to the unexplained claim that they are accidental generalities, and consequently not laws. The conclusion that they fail to be laws also follows on our account.

The most interesting example of the three is H. Reichenbach's example of two statements of the same logical form:

(**U**) All uranium cubes are less than one cubic mile in volume.
(**G**) All gold cubes are less than one cubic mile in volume.

[18]*The Structure of Science,* Harcourt, Brace & World, Inc., 1961, pp. 58–59.
[19]*Philosophy of Natural Science,* Prentice-Hall, Inc., 1966, p. 56.

The standard assessment is that (**U**) is a law and that (**G**) is not.

One reason given for (**U**) being a law is that it follows from the laws of Quantum Mechanics, while (**G**) is usually considered to be obviously accidental.[20] The mini-theory (**LAGT**) doesn't yield the conclusion that (**U**) is a law (The mini-theory does not provide sufficient conditions for being a law). Nevertheless it does yield the important result that (**U**) is not accidental. The reason is that there is a theory, namely Quantum Theory, that explains all the instances of (**U**), and that is enough, as we have seen, to guarantee that it is not an accidental regularity.

To see that (**G**) isn't a law, recall that according to (**LAGT**), we have $\mathcal{L}(\mathbf{G}) \Rightarrow \mathbf{LUCT}(\mathbf{G})$. Moreover, according to **LUCT(G)** there is a theory that explains why each cube of gold is less than one cubic mile in volume. But there isn't such a theoretical explanation for all the cubes of gold. So, **LUCT(G)** is false, and consequently, (**G**) is not a law.

The familiar examples that Nagel and Hempel provided as non-laws, (N), and (H), also turn out to be non-laws on our account. The argument is the same as for (G). Our account requires that $\mathcal{L}(\mathrm{N}) \Rightarrow \mathrm{LUCT}(\mathrm{N})$. Now LUCT(N) is false since there isn't any theory that explains why each of the screws in Smith's car is rusty. Similar verdicts hold for Hempel's "All the rocks in this box contain iron." There's no theory that explains why all those rocks contain iron.

In laying out the claims of both (LAG) and (LAGT), we do not by any means think that they are all that can or should be told about laws and accidental generalizations. We think they are just part of a fuller story.

It is true that on the basis of this account so far, if a generalization is explained, then it is not accidental. The obvious corollary is that there is no explaining an accidental generality. It must be stressed however that that result does not diminish their scientific importance, utility, or even their use in explaining other regularities.

In the preceding chapter we discussed the virtues and defects of accounts of laws that were already very well known, sometimes supplementing them to represent them in their strongest form. In the end we found them unsatisfacory. In contrast to those, we proposed a mini-theory (LAG) that intertwined the notions of law and explanation and believe that the simple consequences were of great interest, and that (**LAG**), even though mini, deserves further development, and consideration.

We turn now to two neglected accounts of the ***explanation*** of laws which, despite the well-deserved repute of their authors, have been overlooked because they have not been seen as the innovative and interesting proposals they are. One account, due to Ernest Nagel we believe to be very radical and reflects the influence of David Hilbert. The other, due to Richard Braithwaite, is a response to Ramsey's critique of laws due to John Stuart Mill. We think that Braithwaite's proposal is a better answer

---

[20]Though it is possible to accept these judgments, the argument that (U) is a law because it follows from laws is not a good argument. It is not generally true that every logical consequence of laws is also a law.

to Ramsey's objections than the well known response by David Lewis. We will discuss both accounts of laws and their explanations by a detailed consideration of the following points in the following chapter:

1. Both Nagel and Braithwaite were influenced by Norman Campbell's account of theories and their laws.
2. Both thought that all explanations, of singular facts were deductive, and that explanations of laws had to be deductive as well. This latter idea is familiar but we will argue that is seriously mistaken. We will argue that some very powerful theories are schematic, and that many laws are cases of *subsumption under schematic theories,* and not cases of deduction.
3. Nagels account, I believe, has to be adjusted to take into account his deep indebtedness to Hilbert's work in Geometry. Nagel's account of theories, is, according to our reconsruction of it, much more radical and challenging than that has been thought.
4. Braithwaite was among the first to give an account of laws and their explanations. It was a reply to Ramsey's objections to the Millean account of theories that was, in my opinion, even better than the one now known as D. Lewis' systems account of laws. These two theories have a host of very interesting consequences, which we will discuss in the following chapter.

**Acknowledgements** I am grateful to Professor Alberto Cordero for his enthusiasm and interest about the central idea of this essay early on, to Professor Hugh Mellor, to members of my CUNY Graduate Center seminar on scientific laws, to the audience of a 2014 conference of the Mind Society and the Aristotelian Association in Cambridge UK, and to two anonymous referees for helpful comments., and published in *Conceptual Carifications, Tributes to Patrick Suppes (1922–2014)*, D. Gabbay, College Publications, Co. Uk, 2015.

## References

Braithwaite, R. B. (1953). *Scientific explanation*. New York: Cambridge University Press.

Carroll, J. (2008). Nailed to hume's cross. In J. Hawthorne & D. Zimmerman (Eds.), *Contemporary debates in metaphysics*. Oxford: Blackwell.

Hempel, P. (1966). *Philosophy of natural science*. Englewood Cliffs: Prentice-Hall.

Hempel, P. (1965). *Aspects of scientific explanation*. New York: The Free Press.

Johnson, W. E. (1964a). *Logic Part I*. Cambridge University Press, 1921, Reprinted by Do-ver Publications, NY.

Johnson, W. E. (1964b). *Logic, Part III*, The logical foundations of science, Reprinted by Dover Publications, NY.

Koslow, A. Explanation and modality, Typescript.

Lotze, H. (1887). *Outlines of logic and of encyclopedia of philosophy* (ed. and trans. G. T. Ladd). Boston: Ginn & Co.

Nagel, E. (1961). *The structure of science*. New York: Harcourt, Brace &World, Inc.

Roberts, J. T. (2008). *The law-governed universe*. Oxford: Oxford University Press.

[illegible] of [illegible] the well known response [illegible] discuss both [illegible] their explanations by [illegible] the following points in the [illegible] chapter.

1. Both [illegible] were [illegible]
2. Both [illegible] deductive [illegible] explanation [illegible] will [illegible] theories [illegible]
3. [illegible]
4. [illegible] the [illegible] chapter.

[illegible]

## References

[illegible]

# Part II
# Theoretical Scenarios, Schematic Theories and Physical and Nomic Modals

# Chapter 8
# The Explanation of Laws: Two Neglected and Radically Different Theories. One Inspired by D. Hilbert; the Other Inspired by F. Ramsey

> *Even if we were sure that all possible laws had been found and that all the external world of nature had been completely ordered, there would still remain much to be done. We should want to* explain *the laws*
> N. Campbell, What is Science, Methuen & Co. Ltd., London (1921), p.77.

Norman Campbell rightly set the task. It is the business of science not only to discover laws, but to explain them. And he added his voice to a philosophical tradition going back to Aristotle, of taking on the task of explaining what laws are, and explaining as well what explanations of laws are. Ever since the renewed interest spurred by seminal paper of Hempel and Oppenheim on scientific explanation,[1] philosophers have been inspired to do better on scientific explanation. But it became painfully clear, from the counter-example offered in their paper, that their account of scientific explanation could not cover the explanation of laws. It is clear that although this is the business of philosophers, it is still unfinished business.

## 8.1 The Campbellian Background

Ernest Nagel, and Richard Bevan Braithwaite were well aware of Campbell's views on the structure of theories, and referred to them when they addressed the issue of the explanation of laws directly. Both had, as we shall see, very different accounts of

This chapter is based upon Koslow, A., "The explanation of laws: some unfinished business", *The Journal of Philosophy*, Special Issue: Aspects of Explanation, Theory, and Uncertainty: Essays in Honor off Ernest Nagel, 2012, pp. 479–502.

[1]"Studies in the Logic of Explanation", Philosophy of Science, 15 (1948), pp.135–178.



A. Koslow, *Laws and Explanations: Theories and Modal Possibilities*, Synthese Library 410, https://doi.org/10.1007/978-3-030-18846-7_8

laws, and their explanations. Braithwaite (1953) developed a view that can be traced back to J.S. Mill and F.P. Ramsey (a view which Ramsey later rejected) that was a very different variation of the Mill-Ramsey view that David Lewis developed some two decades later (1973). Nagel developed a novel view that is an interesting combination of the views of Norman Campbell and David Hilbert. Sad to say, however, both of these remarkable accounts never got the timely critical attention that they deserved. In our view both accounts were mistakenly regarded as just minor variations of a "received view" that was proposed by Hempel. We think that both accounts were strikingly radical for their time, and perhaps even for ours.

## 8.2 Deductive Systems and Their Laws

In a very laudatory review of Richard Braithwaite's *Scientific Explanation*[2] Ernest Nagel raised an objection to Braithwaite's account of scientific laws which is worth quoting in full:

> Braithwaite's explicit formulations of the structure of explanatory systems . . . do not always make entirely clear whether he thinks that the deduction of an empirical generalization from some higher-level hypothesis is sufficient to constitute an explanation for the generalization- . . . In any event, it is doubtful whether most physicists would accept as an "explanation" of, say Galileo's law for freely-falling bodies ($d = gt^2$) the derivation of this law from the logically equivalent hypothesis that $t = \sqrt{(2dg)}$ . Deducibility from higher-level hypotheses is at best only a necessary condition for explanation, not a sufficient one. The explanatory hypotheses in most of the actual scientific systems are required to meet other conditions as well: possessing a greater "generality" (not to be identified with greater deductive power) than the lower-level hypotheses.[3] . . . It would improve his exposition, nevertheless had he discussed them systematically and so guarded himself against the possible misconception that the sole task in the quest for scientific explanations is the relatively easy one of constructing a deductive system in which empirical generalizations are theorems. (205).[4]

The example of two logically equivalent versions of Galileo's law of freely falling bodies is beside the point. In the kind of scientific deductive system that Braithwaite considered (a novel kind of structure we shall try to explain below), it would not be acceptable to have an hypothesis be a higher-order hypothesis, and its logical equivalent be a lower-order hypothesis in the same deductive system. Nevertheless Nagel's point is a good one. It does seem that typical examples of an explanations of laws involve a least a premise of greater generality.

The second admonition by Nagel is that there is the impression –to be guarded against- that explanation of laws amounts to showing them to be theorems in a

[2]Cambridge University Press, 1953. All page references for Braithwaite are to this book.

[3]We omit the other features Nagel mentioned since they concern analogy and evidential requirements that are not directly relevant to the issues under discussion.

[4]"A Budget Of Problems In The Philosophy of Science.", *Philosophical Review*, 66(1957), 205–225.

deductive system. He thought that providing explanations, in that case, would be a relatively easy task. However Braithwaite's proposal for distinguishing the *laws* of a well established deductive system is simply this:

> The condition for an established hypothesis *h* being *lawlike* (i.e., if true, a natural law) will then be that the hypothesis either occurs in an established scientific deductive system as a higher-level hypothesis containing theoretical concepts or that it occurs in an established scientific deductive system as a deduction from higher-level hypotheses which are supported by empirical evidence which is not direct evidence for *h* itself. (*Scientific Explanation*, 301–302)

It's not so evident that such *established* deductive systems are relatively easy to come by. From this passage and others, it is clear that Braithwaite had proposed conditions for something to be a law. His conditions for something to be an explanation of a law are another matter entirely. Nagel, however took these conditions to be those for an explanation of a law, rather than those for being a law.

As I understand Braithwaite's project, the idea is to single out those generalities of an established deductive system which are its laws. His proposal could be stated this way:

> Let S be some generalization of some established deductive system D. Then S is a law of D if and only if there is some generalization S$*$ of D which implies S (but not conversely).[5]

The place of a generalization *in a deductive system* is the key ingredient in sorting out the laws from the other generalizations of the system.

Braithwaite's reliance on the place that a generalization has in a deductive system is an unusual requirement. The usual accounts of laws usually make no reference to position in a structured body of statements. For example, there is the requirement that laws are those contingent generalizations that are not accidental. There is also the claim that laws are those generalizations that support their corresponding counterfactuals. Then too, there is the account according to which laws are those contingent generalizations that are physically necessary. None of the notions of "accidental generalization", "counterfactual conditional", or of "physical necessity" appeal to some structured body of statements.[6]

[5]We have added the parenthetical condition to avoid the consequence that every generalization of an established deductive system would be a law. It still does not blunt the possible case where a deductively strongest generalization of a deductive system might not be a law of that system. This is a consequence which Braithwaite called to attention, and developed an answer. It should also be noted that although such a highest generalization might not count as a law of a system D, it might very well be a law of an established deductive system that was an extension of D. His additional requirement that a highest-order hypothesis in a deductive system that has an occurrence of a theoretical term is a law seems very ad hoc to me. Nevertheless there are several important plausible cases when this is so –in systems in which the highest order hypothesis are the three laws of Newtonian mechanics, Newton's theory of gravitation, Schrödinger's formulation of quantum mechanics, and the basic (three or four) laws of classical thermodynamics.

[6]I am not endorsing any of these accounts. In fact there are fairly convincing examples of accidental generalizations that also support their corresponding counterfactual conditionals. Even worse, we showed in Chap. 3, that if a law implies its corresponding counterfactual conditional, then it is

The explanation of laws is another closely related matter which we will examine in detail later in this essay. We shall simply for the present call attention to two seminal paragraphs that capture Braithwaite's views on explanations of laws.

(1) "To explain a law, as we have seen, is to incorporate it in an established deductive system in which it is deducible from higher-level laws. To explain these higher-level laws is similarly to incorporate them and the deductive system in which they serve as premises, in an established system which is more comprehensive and in which these laws appear as conclusions. To explain the still-higher-level laws will require their deduction from laws at a still higher level in a still more comprehensive system."(347), and
(2) "Any incorporation of a fact – be it a particular instance of a law or the law itself – into a deductive system in which it appears as a conclusion from other known laws is, by virtue of that incorporation, an explanation of that fact or law. ... what matters is that we know more than we did before of the connectedness of the fact or law with more fundamental laws covering a wider range." (349)

It is however certainly true, as Nagel noted, that an explicit account of one law being more general or more fundamental, or having wider range than another, is patently missing from this account of the explanation of laws. This we shall try to rectify by a slight addition.

Braithwaite's account of the explanation of laws in (1) and (2) relies upon the use of established deductive systems and is very different from his account of laws. We turn then to his account of laws, and then to his account of their explanation, after some remarks about his notion of a deductive system.

## 8.3 Deductive Systems and Theories

It is not at all obvious what Braithwaite meant by "deductive system". Although influenced by Campbell's account of theories which we shall describe more fully below, deductive systems are not theories in Campbell's sense. Here is Braithwaite's simplified example, – a "Gallilean"deductive system, based roughly on Galileo's law of falling bodies. (13) Informally the statements in the system are arranged in levels with those on the higher levels implying (but not being implied by) those on the lower. The idea was that

> The hypotheses in this deductive system are empirical general propositions with diminishing generality. (13)

The system he had in mind is illustrated by the following statements in descending order of generality;

---

equivalent to that counterfactual. In that case there are examples of laws which scientific practice regards as logically equivalent, while their corresponding counterfactuals are not.

(I) Every body near the earth freely falling towards the Earth falls with an acceleration of 32 feet per second per second.
(II) Every body starting from rest and freely falling towards the Earth falls $16t^2$ feet in $t$ seconds, whatever number $t$ may be.
(IIIa) Every body starting from rest and freely falling for 1 second towards the Earth falls a distance of 16 feet. (And so on for other values of $t$).

It is obvious from this example that a deductive system cannot be identified with a set of axioms from which all the other statements in it follow (the elements in (IIIa) are not axioms of the system). Nor is it the set of all logical consequences of a set of axioms (logical truths are not in it). So deductive systems are neither a theory in the sense of a set of axioms like the axioms for Euclidean geometry, nor are they a theory in the sense of Tarski (the set of logical consequences of some set of sentences).

Probably one way to represent Braithwaite's deductive systems would be as a lattice (not Boolean, since that would restrict systems to those having exactly one higher-order hypothesis, which would be too restrictive). A lattice structure would allow one to place hypotheses of various generality at various nodes with arrows indicating a type of implication.

A more formal description of these deductive systems is not a serious problem. Here is one suggestion: what is needed is a notion of a structure that contains more than the listed axioms of a theory in the first sense of "theory" we discussed, and less than a full Tarskian theory. We shall define a *Braithwaitean deductive system* D$*$ as any set of statements which satisfies the following conditions, where S and S$*$ are any contingent hypotheses:

(1) D$*$ contains all the axioms of some theory T,[7]
(2) If S logically implies S$*$ (S $\Rightarrow$ S$*$), then if S is in D$*$, so too is S$*$.
(3) If S is in D$*$, and I is an instance of S, then I is in D$*$,

where (2) is a kind of closure condition restricted to contingent generalizations.

It is clear however that although the structure of the system involves deductions, it also involves levels of generality, although an account of "is more general than" is missing. We shall see below that it was Nagel who offered an explicit account, – actually two - of "is more general than", one more general than the other (no pun intended).

Both Braithwaite and Nagel thought that something like greater generality had to figure in the explanation of laws. The requirement, as Nagel expressed it, is that for the explanation of laws, at least one of the premises has to be more general than what is explained. Of course the requirement doesn't seem to be needed for explanations of singular facts. Neither does it figure in those explanations of events that consist in locating them in a causal network. The fact that some notion of generality figures in explanations of laws but not in other kinds of explanations is worth worrying about –

[7]We believe that here Braithwaite would have used something close to Campbell's notions of the hypothesis and the dictionary of a theory . Those notions will be discussed more fully below.

and we will tentatively offer a mini-theory at the end of Chap. 12, that that connects the generality of theories to those laws that they explain.[8]

As we noted above, the account of laws that situates or relativizes them to deductive systems, is unusual. This use of deductive systems has an interesting history, and before we turn to Braithwaite's account of the explanation of laws, it is important to consider the problem that Braithwaite attempted to solve with that account.

## 8.4 The Mill-Ramsey-Braithwaite Account of Laws (MRB)

The origin of the Mill-Ramsey-Lewis (**MRL**) account of laws begins with some remarks of the Cambridge philosopher W.E. Johnson about two kinds of conditionals – universals of fact, and universals of law. That prompted a sharp reply by another Cambridge Philosopher, F. P. Ramsey. Ramsey then entertained an account that essentially had been proposed earlier by J.S. Mill – the Mill-Ramsey account (**MR**). Soon thereafter, Ramsey (and, I am fairly sure, Braithwaite) rejected (**MR**). Braithwaite then developed a better version of it. Some two decades after that, David Lewis developed a different non-epistemic version (**MRB**) of the discarded Mill-Ramsey account. It is the remarkable but neglected Braithwaite account that we will now describe.

Ramsey's sharp reaction to Johnson's distinction between two kinds of conditionals was that it was impossible. Johnson thought that universals of fact were universal quantifications over all things that in fact satisfied a conditional, whereas with universals of law, the quantifier ranged over all possible things that satisfied a conditional. Ramsey (an unpublished note) remarked that Johnson got the quantification wrong and failed to realize that "everything" means everything. Ramsey then considered an account of laws that J.S. Mill had advocated:

> ... What are laws of nature? May be stated thus: What are the fewest and simplest assumptions, which being granted, the whole existing order of nature would result? Another mode of stating the question would be this: What are the fewest general propositions from which all the uniformities which exist in the universe might be deductively inferred?.[9]

Ramsey's compact expression of that view was simply that

> Laws are consequences of those propositions which we should take as axioms if we knew everything and organized it as simply as possible in a deductive system.[10]

[8] I suspect that part of the reason is that "is more general than" was sometimes used interchangeably with "is more comprehensive than" and with "is more fundamental than".

[9] J.S. Mill, A System of Logic, 1843, p.230

[10] F.P. Ramsey, Philosophical Papers, ed. D. H. Mellor, Cambridge University Press, 1990, p.150.

However Ramsey (and I am fairly sure Braithwaite) rejected it for two reasons: that it is impossible to know everything and also to organize it in a deductive system.

The Millean proposal then was developed in at least three different ways. Ramsey went on to consider laws as variable hypotheticals – a device which isn't even propositional. I shall say no more about this possibility. Another better known response is due to David Lewis who expunged the epistemological element from Ramsey's formulation, and replaced the deductive system of the totality of everything known, by another deductive system, and accounted for laws this way:

> ...a contingent generalization is a law of nature if and only if it appears as a theorem (or axiom) in each of the deductive systems that achieves a best combination of simplicity and strength[11]

With this variation on the Mill-Ramsey theme we have one of the most durable and plausible Humean accounts of law in contemporary philosophy of science – (**MRL**).

With the Mill-Ramsey version (**MR**), we had a deductive system, some of whose propositions were known to be in it, but certainly not all. With Lewis' variation, we have an appeal to a deductive system for which it is not a sure thing that one's favorite generalizations will be included in it. That is, of all the true deductive systems that are possible, the laws will be members of the one which is the best combination of simplicity and informativeness. Even if it were settled how simplicity and informativeness were to be construed, it is less than satisfying if the answer to the question "Is P a law?" for any true generalization P, is: "I don't know", "I'm not sure or "It's anyone's best guess." The epistemic reservations raised by Ramsey remain in force.

Braithwaite's proposal (**MRB**) is a variation on the Mill-Ramsey account that is significantly different from the variation proposed some two decades later by David Lewis. Lewis expunged the epistemic drawback of the Mill-Ramsey proposal. Braithwaite's idea was to meet, in one fell swoop, the two objections that Ramsey raised: (1) that it is impossible to know everything and (2) impossible also to organize to organize everything known into a deductive system. Instead of one axiomatization of all our knowledge (as D. Lewis proposed), to use instead various *established* deductive systems, each organized organized around specific Campbellian theories.[12]

---

[11]D. Lewis, Counterfactuals, 1973, p.73. A more nuanced version can be found in his "Humean Supervenience Debugged.", *Mind,* 1994 (473–490).

[12]One interesting difference between (MR) and (MRL) on the one hand, and (MRB) on the other, is that all the statements of the first two kinds of deductive systems are required to be true, but the statements of (MRB) only need to be well established.

## 8.5 Deductive Systems and the Explanation of Laws

If we turn to (1), the first of the two paragraphs in which Braithwaite explicitly described the condition for an explanation of a law, we find an intricate use of deductive systems and their proper extensions. Paragraph, (1), can be described as a compact condition on explanations this way:

> (**E**) Let A be a law in the deductive system D. A is explained by L if and only if L is a law in some deductive system D∗ that is a proper extension of D, and L implies, but is not equivalent to A.

Basically the idea is that you cannot explain a law in a deductive system by restricting yourself to the members of that system. You have to use a law (or laws) in a proper extension from which it follows. One of the best examples of explanation (one of several) for him was the explanation of an approximate formulation of Newton's law of Gravitation by Einstein's General Theory of Relativity.

In (2), the second paragraph, (2), Braithwaite wrote that the explanation of a law is the incorporation of it into a deductive system. We take him to mean that it is a case of the incorporation (we prefer to say "the embedding") of one entire deductive system within another, *where the explaining law is not a member of the embedded system*. Braithwaite thought that the larger system had greater predictive power. To that extent he probably thought of the larger system as more comprehensive. However that systemic consideration does not lead, as far as I can see, to any notion of one law being more general than another. Nagel's condition that any explanation of a law requires the use of some more general law is not vindicated on the Braithwaite account. Nevertheless it has some significant features that we can only mention at present:

(1) It does not fall victim to the Hempel-Oppenheim counter-example which they raised against their own Deductive Nomological Model, when applied to laws. The Hempel-Oppenheim model would require that the conjunction of Kepler's laws of planetary motion and Boyle's law of gases explains Kepler's laws. It can be shown that that absurd example is not permitted according to Braithwaite's proposal

With Braithwaite-style deductive systems in place, there are also some nice consequences that should be noted:

(2) (**LL**) If L is a law of a deductive system D, and A is a contingent generalization) that is logically implied by L, then A is also a law of D.

I.e., contingent generalizations that follow from laws of D, are also laws of D.[13] So the prefix "It is a law of a deductive system that ... "distributes over implication.

[13] The more general result also holds: If L∗ is a generalization that follows from several laws of D, then it too is a law of D.

(3) **(EL)** If A is a contingent generalization of D that is explained in some deductive system D∗ then A is also a law of D∗.

(3) provides a rationale for seeking explanations of contingent generalizations. If you think that laws are important, then explanations are one way of assuring that explained generalizations will also be laws. Lastly, there is a consequence, similar to (EL), for explanations of laws.

(4) **(EE)** If A is a contingent generalization in the deductive system D that is explained in the deductive system D∗, and A∗ is a contingent generalization implied by, but not equivalent to A, then A∗ is also explained in D∗.

The Braithwaite variation on the Mill-Ramsey proposal hasn't been recognized for what it is – a Humean account of laws and their explanations, that exploits the relation that laws have to those established deductive systems that contain them, and connects the explanation of laws to the embedding of established deductive systems within more comprehensive ones. The connection with scientific practice is evident and we believe the formal features it has, heightens its interest. We believe that it is a step in the right direction. Unfortunately, there are theories which yield laws, but not deductively. They are schematic theories which we will discuss in the next Chap. 9.

## 8.6 Ernest Nagel's Theory; Campbellian Origins

Nagel's account of the explanation of laws is deeply indebted to Campbell, especially for the novel way in which the Campbellian representation of theories is deployed to yield a radical account of the explanation of laws.

Nagel thought that all explanations of laws were deductive.[14](33) Drawing upon a detailed example of an explanation of the law that ice floats on water, here is his compact description of at least three conditions that would hold for any explanation of laws (including statistical ones):

> . . . all the premises are universal statements; there is more than one premise, each of which is essential in the derivation of the explicandum; and the premises taken singly or conjointly, do not follow logically from the explicandum. (34)

Two caveats: We will not be concerned with Nagel's discussion of the explanation of statistical laws.[15] We will also separate the cases when the explanation of a law involves only "empirical" laws as premises, (they contain no occurrences of

[14]All page references are to Ernest Nagel, *The Structure of Science, Problems in the Logic of Scientific Explanation*, Harcourt, Brace & World, Inc., 1961.

[15]His conclusion, after a review of typical examples, is that explanations of statistical laws are deductive, at least one premise is statistical, and at least one premise must have a greater degree of statistical dependence than that of the law to be explained.(520)

theoretical terms) from those in which at least one "theoretical" assumption is used.[16] Nagel assumed that it made no difference, – though, as we shall explain below, we think it does.

With some differences, Nagel adopted Campbell's canonical representation of theories as given by two mutually exclusive sets of statements. The first set, the *hypothesis* of the theory, contains those statements of the theory whose only occurrences of non-logical terms are theoretical. The second set, the *dictionary* of the theory, contains those statements that have occurrences of both theoretical and observational terms. It is important to note that for Campbell, the hypothesis and the dictionary consist of statements which have a truth-value. Nagel adopted this canonical description of theories with one very critical difference. He associated three components with each theory:

(1) an abstract calculus that is the logical skeleton of the explanatory system, and that "implicitly defines" the basic notions of the system.
(2) a set of rules that in effect assign an empirical content to the abstract calculus by relating it to the concrete materials of observation and experiment, and
(3) an interpretation or model for the abstract calculus, which supplies some flesh for the skeletal structure in terms of more or less familiar conceptual or visualizable materials. (90)

Clearly, Nagel intended not only to reflect, but to improve upon Campbell's canonical representation of how theories explain laws.

For Campbell, the representation of the required deduction would look roughly like this: (1) a finite number of hypotheses relating only some of the theoretical terms to other theoretical terms), (2) a finite number of dictionary entries which Nagel called "coordinating definitions" (Hans Reichenbach's terminology) relating some of the theoretical terms to some observational ones, and finally (3) the logically deduced conclusion, – some law L.

## 8.7 Explanation and Theories Without Truth-Values

There is no point in trying to indicate the forms that the sentences in each group might have in order to insure that the deduction is correct. There are too many theories with various forms to do that. Nagel's representation of the explanation of a law would have these features: the premises would include some members of the abstract calculus of the theory, together with some statements that correspond to what Campbell called the dictionary of the theory, followed logically by the conclusion –the law to be explained. Nagel calls the dictionary items rules, relating

---

[16] Nagel was certainly aware of the various criticisms of such a distinction between theoretical and observational terms, but he has a very vigorous account of that distinction, admitting its vagueness, that should not be discounted.

theoretical to observational terms, but I don't think it matters much for the following argument whether we use the statement-version or the rule-version for the dictionary entries. So, roughly, Nagel's representation of the explanation of a law should look something like this simple example (modulo the point that most everyone agrees upon, that in such explanations there would usually be two hypotheses):

(i) T $(\tau_1, \tau_2)$ (a theoretical statement)
(ii) D$(\tau_1, o_1)$ (a dictionary entry)
(iii) L (the law to be explained)

By focusing on Nagel's construal of (i) and (ii), it will become apparent that Nagel had in fact proposed a bold variation on Campbell, that resonates with a tradition that goes back at least to the geometer Moritz Pasch, and more clearly to a notion of axiomatization that David Hilbert advocated for any theory (mathematical or physical). The geometric studies of both mathematicians were well known to Nagel.

On Campbell's account, (i) is an hypothesis relating two theoretical terms, and hypotheses are statements that have a truth-value. In contrast, Nagel thought that (i) was not a statement at all, and was neither true nor false. Campbell regarded the dictionary entry (ii) of the theory as a statement that had a truth-value. In Nagel's version, the coordinating definitions are now described as rules. For that reason, they too are neither true nor false.[17]

Nagel's reason for taking Campbell's hypotheses as lacking truth value is simple enough, though Nagel never made the argument explicit. I believe it rested on two assumptions. The first is that he thought that an "abstract calculus" (his version of Campbell's hypotheses of a theory) was an axiomatization of the theory, and second, he also thought that axiomatizations of theories were to be understood in a way that Hilbert made famous with his special concept of axiomatization in mathematics and the sciences. That, view, as we shall see (in Chap. 9), involved understanding that the axioms of theories failed to be statements, and thus were neither true, nor false.

## 8.8 Explanation: Axiomatization and the Hilbert Connection

Although Nagel didn't argue for these assumptions, the reasons he held them seem to me to be evident. If we look to the examples that he provided for those theories he called "abstract calculi", three are singled out: Euclidean geometry, the kinetic theory of gases, and probability theory as axiomatized by A. Kolmogorov. Concerning Euclidean geometry, Nagel said that

[17]Consequently, (ii) should be listed as a rule, rather than as a premise. Whatever status it has, it is supposed to guarantee a deductive transition to the conclusion, since, for Nagel, Campbell and Braithwaite, as we indicated, all explanations are deductive.

> The postulates of the system are frequently stated with the expressions 'point', 'line', plane, 'lies between', 'congruent with', and several others as the basic terms. . . . Indeed, in order to prevent the familiar although vague meanings of those expressions from compromising the rigor of proofs in the system, the postulates of demonstrative geometry are often formulated by using what are in effect predicate variables like 'P' and 'L' instead of the more suggestive but also more distracting descriptive predicates 'point' and 'line'. (91–92)

The example of probability theory is especially interesting because it makes evident the connection of Nagel's view of these axiomatizations with Hilbert's axiomatization of Euclidean geometry. Kolmogorov said explicitly that in providing axioms for probability, he was trying to do the same for that theory, that Hilbert did for Euclidean geometry. Nagel, as we noted, was acutely aware of the development of this Hilbertian view of the axiomatization of Euclidean geometry and its earlier anticipation by Pasch. In his penetrating early study of the development of geometry[18] he noted:

> Indeed [Pasch declares], if geometry is to be really deductive, the deduction must everywhere be independent of the *meaning* of geometrical concepts, just as it must be independent of the diagrams; only the *relations* specified in the propositions and definitions employed may legitimately be taken into account. During the deduction it is useful and legitimate, but in *no* way necessary, to think of the meanings of the terms; in fact, if it is necessary to do so, the inadequacy of the proof is made manifest. If however, a theorem is rigorously derived from a set of propositions – the *basic* set- the deduction has a value which goes beyond its original purpose. For if, on replacing the geometric terms in the basic set of propositions by certain other terms, true propositions are obtained, then the corresponding replacements may be made in the theorem; in this way we obtain new theorems as consequences of the altered basic propositions without having to repeat the proof. (237–238)

This insight is incorporated into the very heart of Nagel's account of explanation of laws by theories. The upshot of these considerations was to replace terms like "point" and "line" in a theory of Euclidean geometry by predicate variables –say "P" and "L"". This rewrite has the result that (i) which looked like a statement having a truth-value has now been replaced by what Nagel called a statement-form. The idea is to do the same for the basic predicates and relations of the kinetic theory of gases, and probability theory. Something quite radical then results when this Hilbertian insight is introduced into the original Campellian representation of theories.

Consider, for example, what happens to the explanation of a law, in which the deduction is given by the proof-sequence (i) to (iii), once we replace (i) and (ii) by the statement-forms (i)′ and (ii)′. The explanatory proof then is given by

(i)′ T (X1, X2)
(ii)′ D(X1, o1)
(iii) L,

[18]"The Formation of Modern Conceptions of Formal Logic in the Development of Geometry", *Osiris*, vol.7, 1939,and reprinted in E. Nagel, *Teleology Revisited and Other Essays in the Philosophy and History of Science*, Columbia University Press, 1979, pp.237–8. The particular passage (translated by Nagel) is also reprinted in P. Suppes, *Representation and Invariance of Scientific Structures,* CSLI Publications, Stanford, 2002, p.46 where it is used to support a set-theoretical account of axiomatic theories.

where the theoretical terms have been replaced by the predicate variables X1, and X2. The results are radical, and initially implausible. The first thing to note is that the deduction fails to be an explanation of L. The reason is that explanations are factive – that is, the sentences that are used in explanations (whether as premises, or conclusions) are true.[19] Here however, "T (X1, X2)" is, according to Nagel, a linguistic expression that has predicate variables in the place where the usual representation had specific theoretical terms. I have also replaced the theoretical term $\tau_1$ in (ii) by a predicate variable. That might be somewhat unfair to what Nagel intended. Some terms in a theory may not have some experimental notion associated with them (they don't occur in the dictionary). He says "in effect those terms have the status of *variables*." (132, emphasis E.N.) So it looks like the occurrence of the theoretical term in (ii) should not be replaced by a predicate variable in (ii)' since $\tau_1$ has an experimental notion associated with it. Nevertheless, if the theoretical terms in the hypothesis part of the theory are replaced by variables, and if those same terms which may also occur in a dictionary entry are not replaced by a variable, but are kept as they are, then the deductive connection may be broken, and the explanation destroyed. Here's an example of that possibility (assuming that $\tau$ and $\mu$ are monadic theoretical terms, o and o' are experimental ones, and the universal quantifiers are suppressed):

(iv) $\tau = \mu$
(v) $o \rightarrow \tau$
(vi) $\mu \rightarrow o'$
(vii) $o \rightarrow o'$

This is a perfectly fine deduction of what might be an experimental generality (vii). However if we replace the terms in (iv) by distinct variables, but leave the occurrences of those terms untouched in the dictionary entries (v) and (vi), the deduction of (vii) is ruined. It is obvious that in order to retain the deduction, you either have to leave all the theoretical terms alone, or replace them all by predicate variables. Since it is a deductive explanation that is at stake, I think the thing to do is to replace the occurrences of the theoretical terms by appropriate variables. In that case, we are left with the task of trying to explain the rules provided by (v) and (vi). The original intention of dictionary entries was to show how a theory can be related to experimental laws, but it is something of a mystery to me how that is achieved by having the dictionary give various ways in which observational terms are related to variables. Thus, if the theoretical statements have no truth-value, I think the same thing goes for the dictionary entries. Although Nagel thought of entries like (v) and (vi) as rules, I do not see how that removes the difficulty.

[19] Nagel's view is subtle. He does say in effect that *if* it is required that every premise in an explanation is either true or false, then it is almost unavoidable to require that they be true (42–43). Factivity for him is a conditional. So, for him one could say that factivity holds (vacuously) for those premises in explanations that are, like (1)', without any truth-value.

There is another difficulty, relatively minor, with Nagel's proposal to replace theoretical terms with variables. It treats variables in a very non-standard way. Normally, if we have a variable ranging over a certain domain, and want to consider a special case, then the usual thing is to use the name of the special element of the domain. So if one wanted to go from "x is a prime number other than 2, and is odd" which has no truth value, to the special case of the number 3, one would use the name of that number to obtain "3 is prime number and is odd." Nagel however used another convention. His example (132) is

(1) For any x, if x is an animal and x is P, then x is a vertebrate.",

where 'P' is a predicate variable, so that we have an expression that has no truth-value. He suggested that one way to obtain a true statement was to substitute "mammalian" for the predicate variable. The result of this "substitution" is

(2) For any x, if x is an animal and x is mammalian, then x is a vertebrate.

What Nagel says is of course true, but it's not the way substitution for variables goes. If it were a case of substitution, then what would replace the variable is the name of something in the range of that variable. In the case of predicate variables, the substitution requires the name of a predicate and not the predicate itself. Strictly speaking then, the use of predicate variables is clearly being used in a non-standard way. In short, Nagel's construal of theoretical terms as predicate variables is unfortunate. It threatens even his central theme that all explanations are deductive.

## 8.9 Explanation: The Schematic Account

We think that what would better suit Nagel's purpose is a fresh start with the use of some device other than predicate variables to describe those premises that use theoretical terms in the explanations of laws. We suggest the systematic use of schematic letters to construct a new, but closely related account – call it N∗. It is a close cousin of Nagel's account, and would better represent his view as closely aligned with the views of writers like M. Pasch and D. Hilbert.

Schematic letters are linguistic expressions of various kinds: schematic sentential letters, and schematic predicate letters for example. Wherever they occur in expressions, they may be replaced by specific sentences and specific predicates (respectively). As we shall see, many of the difficulties connected with Nagel's replacement of theoretical terms can be resolved when they are treated schematically.

Our suggestion, in short, is that instead of replacing theoretical terms by predicate variables, they be replaced instead by schematic letters of the appropriate kind. Nothing has changed in N∗, so far as truth-values are concerned.

Schemas are any expressions that have schematic letters embedded in them, of whatever kind. The way to convert a schema to an expression that has a truth-value, is to replace the schematic letters in them by specific non-schematic expressions of the appropriate kind, and this of course can be done in many ways. What I am suggesting then is that Nagel's view of the abstract calculi that he saw "embedded", is best represented as a schematic theory.

Replacing some of the theoretical terms by the appropriate schematic letters avoids the problems raised by Nagel's use of predicate variables. With a shift to the schematic account N∗, several benefits are automatic. First, the sequence (iv) – (vii) with the theoretical terms replaced by schematic letters, now becomes a genuine deduction –something not true if those terms are replaced by predicate variables.[20] Second, it is impossible to make sense of a dictionary entry that has the form of $o \rightarrow \tau$ (to use the made-up example of ((iv) – (vii) above), if "$\tau$" is replaced by a variable, since there is no sense provided for a conditional with a variable as consequent (or antecedent). However it is relatively easy to make sense of a conditional with a consequent (or antecedent) that is a schematic letter.

As Nagel has emphasized, the dictionary items are there to provide a deductive link to observational material. This, the shift to schematic letters achieves.

## 8.10 Explanation: Whither Campbell's Dictionary?

There is however to my mind a serious oversight with Nagel's claim that if there is to be a link from a theory to observational matters then dictionary items are indispensible. It needs qualification. The claim is plausible if the link is supposed to be deductive.

However, there are two considerations under which the claim is incorrect. The first is that there is a link between some theories and experimental matters, -only that link is not deductive, and not probabilistic either. There are many examples of significant theories that are regarded as having great scientific worth because of their many successes (e.g. The Germ Theory of Disease, the Kinetic Theory of Gases, The Hamiltonian Theory of Least Action are a few examples). These theories do not have the form of universally quantified conditionals. They are "theories-for" rather than "theories-of".[21] They have many experimental successes to their credit, all of which are gathered under them as special cases. None of the successes of these theories is a deductive (or probabilistic) consequence of them. Moreover, no special link or coordinating definition is required to make the connection of those theories with their successes.

The second is that in the present case where the theory is understood to be schematic, there is a way in which a theory can have links to observational matters, without the need to have dictionary entries. Since a properly axiomatized theory in

[20] According to Nagel, "If the theory is to be used as an instrument of explanation and prediction, it must somehow be linked with observable materials." (93).

[21] A discussion of theories-for, theories-of and their importance for understanding the non-deductive relation of them to their successes, and the kind of explanation that they provide, can be found in S. Morgenbesser and A.Koslow, "Theories and their Worth", *The Journal of Philosophy*, CVII, 12, Dec. 2010, pp.615–647.

general (for Hilbert) is schematic, there are many different ways in which the various schematic letters in it can be replaced directly by specific predicates and relations to yield statements, which unlike the schemata, have a truth-value. Let's call the result of such replacements, if true, *applications* of the theory.

Thus, to replace 'point', 'line' and 'plane' now thought of as schematic letters, by 'tables', 'chairs', and 'beer mugs' we do not need coordinating definitions relating these two groups of terms. It suffices to replace one set of terms for the other directly into the schematic theory, and to check whether the result is true. Hilbert himself used such replacements. He noted that by such replacements, some of the axioms of Eudoxus' theory of ratios yielded a general statement about genetics. Of course given his claim that "point" in his axiomatization of Euclidean (plane) geometry could be replaced by "ordered pair of real numbers", and "line" by "linear equation" to obtain a true application, it becomes clear why he wanted to say that ordered pairs of real numbers are also points. It is also clear that he did not mean that points could be explicitly defined as ordered pairs (or triples) of real numbers. Clearly some applications have more interest than others.[22] But none of them are deductive consequences of the schematic theory.

I have no idea how to work out the details of Hilbert's suggestion that 'points', 'straight lines and 'planes' of Euclidean geometry could be replaced by 'tables', 'chairs', and 'beer mugs'(or, as he said in his correspondence with Frege, "instead of points, think of a system of love, law, chimney sweep…which satisfies all the axioms… "However there are other applications like the replacement of 'point' by "intersection of two light rays", and "straight line" by "path of a light ray in homogeneous media" and one can see how the resulting application would be a generalization belonging to geometrical optics, with no need for coordinating definitions.

## 8.11 Explanation, Schematic Subsumption, and Factivity

We shall postpone a more detailed exposition of the features of schematic theories for Chaps. 9, and 12. The detailed examination of this topic, is, we think worth pursuing. It explores how some laws (in the last case, laws like refraction, and reflection in geometrical optics) might be subsumed under a theory. We shall say that the applications, the laws, are *schematically subsumed* under the theory, but are not

---

[22]In fact, Hilbert had a short proof showing that there are an infinite number of such applications, which was subsequently rediscovered by B. Russell, and D. Davidson. ("Any theory can always be applied to infinitely many systems of basic elements. For one only needs to apply a reversible one-one transformation and then lay it down that the axioms shall be correspondingly the same for the transformed things (as illustrated in the principle of duality and by my independence proofs.") *Gottlob Frege,Philosophical and Mathematical Correspondence* (Eds. G.Gabriel, H.Hermes, F. Kambartel, C. Thiel, A.Verart, abridged from German edition by B. McGuinness and tr. By H. Kaal., p. 42.

logical consequences of the theory, meaning that it is just a special case of replacement of the various schematic letters of the theory. The applications of a schematic theory are all uniformly subsumed under the same theory, and that I think counts for some type of explanation. Of course it is not the kind of subsumption endorsed by say Hempel when he said

> ... I think that all adequate scientific models and their everyday counterparts claim or presuppose at least implicitly the deductive or inductive subsumability of whatever is to be explained under general laws or theoretical principles. (424–425)[23]

Hempel was concerned with deductive and probabilistic kinds of subsumption. The applications of a schematic theory, however, are neither deductive nor inductive consequences of the theory from which they were obtained. Nevertheless, if it is possible to think of deductive (or inductive) subsumption as explanatory, then why not make the similar claim when all the applications of the theory are uniformly subsumed under it? At the very least it too provides a unification.

There is still the fact that even if we go schematic with the N∗ version of Nagel's original position, both accounts fail to satisfy the factivity condition for explanations of laws. Theoretical statements have no truth-value, whether, as with Nagel it is because theoretical terms are construed as predicate variables, or whether, as in N∗, the theoretical terms are construed as schematic letters.

For Nagel the distinction between experimental laws (containing no occurrences of theoretical terms), and theoretical laws (which do have one or more occurrences of theoretical terms in them) makes a difference in the issue over factivity. There are a few possibilities worth mentioning: (1) An experimental law is explained by other experimental laws. In this case factivity holds. (2) In the case of an explanation of an experimental law by theoretical ones, matters are different. Some schematic premises deductively yield and perhaps even explain a non-schematic generalization. One example of this possibility might be the argument (iv) to (vii) above. Here the experimental law (vii) is true (say) but what explains it is schematic, and so, without truth- value. In such a case we might say that the explanation is *semi-factive,* a condition which requires only that what is explained is true (rather than what explains must also be true). In this case the situation is comparable to that of the factive "knows that ...". (3) Some theoretical laws explain other theoretical laws. In this case there is no factivity at all; not even semi-factivity since the explainers and the explained don't have truth-values. The possibility of the reduction of one schematic theory to another schematic theory may raise problems for Nagel's theory of reduction since he seems to regard theory reduction as a case of explanation of one theory by another.[24] Well, maybe not. Recall that for Nagel, factivity is conditional: If the items in an explanation have any truth-value at all, then they are true. So in this case, factivity would be satisfied vacuously!

---

[23]C.G. Hempel, *Aspects of Scientific Explanation*, The Free Press, New York, (1965).

[24]"Reduction ... is the explanation of a theory or a set of experimental laws established in one area of inquiry, by a theory usually but not invariably formulated for some other domain." Nagel (338)

There is yet another way in which factivity is involved even with theories that are schematic. Such theories have their applications that are obtained by replacement of their schematic letters with appropriate specific predicates and relations. Those applications of a theory will in general not be schematic, and may in turn figure in the explanation of other non-schematic statements. For those explanations that are based on applications that are subsumed under the schematic theory, non-vacuous factivity still prevails.

## 8.12 Explanation and Generality

We turn finally to a problem which both Braithwaite and Nagel recognized as an important, maybe even a crucial element of the explanation of laws. It seems that there is no explanation of a law that does not involve at least one premise that is more general than it. As far as I can tell, only Nagel attempted a formal account of the concept "is more general than". Unfortunately, I do not see at present how his account fits into our present (schematic) account of Nagel's view of the explanation of laws. There are several drawbacks to the Nagelian account of generality.

First there is the restriction of his analysis to only those laws that are best expressed as universally quantified conditionals. There are, as we have noted, some laws and theories of high scientific merit, which do not have that elementary form. Second, it appears, on his account, that whether one law is more general than another depends critically on syntactic form, so that the order of generality between two laws can change if one uses different, but logically equivalent forms of those laws. And finally, it seems that the account is incompatible with simple examples of explanations offered by Nagel.

Nagel has two accounts of generality: one which addresses the problem of when one group of laws (say from physics) is more general than say a group of biological laws, and a simpler version for comparing one law with another. We concentrate on the simpler version. It goes like this: Let L be the law that all As are B, and L∗ be the law that All Cs are D. Then,

"L is more general than L∗ if and only if it is logically true that All Cs are A, and it is not logically true that All As are C." (38).I do not find it so damaging that the laws for which this concept is defined are limited to generalized conditionals. The definition would still be an important achievement even so. The second feature of this definition is that two logically equivalent expressions of a law can differ in generality. That seems to me not only surprising, but unwelcome. Nagel was well aware of this feature. His example is that

(1) "All living organisms are mortal" is more general than
(2) "All human beings are mortal"

on his account, but

(3) "All non-mortals are non-living organisms"

is not more general than (2)"All human beings are mortal", even though (1) and (3) are logically equivalent.

Nagel has an interesting defense to mitigate this result. It depends upon his deep commitment to a pragmatic view of the matter. He thinks that there is "a tacit reference to contexts of use in the formulation of laws" in which the range of application of the law (indicated by the antecedent of the conditional) can shift. His example is that the common use of "Ice floats in water" has as it's range of application cases of ice that are, were, or will be immersed water, and, he says that rarely (if ever) will the range of application be those things that never float in water. He may be right in this observation, but I do not see what the target of an investigation (what happens to ice immersed in water) has to do with the generality of a statement. That matter may be moot, but there is something more serious at stake. This result leaves open the possibility that you may lose an explanation of a law by using one formulation of a law rather than its logical equivalent. Here's the possibility that I have in mind: Suppose that there is an explanation of the form (i)

(i) $A \rightarrow B$, $[\neg C \rightarrow (B \rightarrow \neg A)]$ implies $(A \rightarrow C)$. Now,
(ii) $([\neg C \rightarrow (B \rightarrow \neg A)]$ is logically equivalent to $[(A \wedge B) \rightarrow C]$, therefore
(iii) $A \rightarrow B$, $[(A \wedge B) \rightarrow C]$ implies $(A \rightarrow C)$,

but (iii) is not an explanation. The reason is that neither of the premises has greater generality than the conclusion. If $A \rightarrow B$ has greater generality than $(A \rightarrow C)$, then $(A \rightarrow A)$ is logically necessary, and it is also not logically necessary. Clearly impossible. And if $[(A \wedge B) \rightarrow C]$ has greater generality than $(A \rightarrow C)$, then $(A \rightarrow (A \wedge B))$ is logically necessary, and $((A \wedge B) \rightarrow A)$ is not logically necessary. Also clearly impossible. Consequently, there can be a loss in explanations with the substitution of logical equivalents.

The most serious drawback to this account of "is more general than", is that there are clear examples of explanations of laws fail to meet the condition that at least one of the premises has greater generality than the law to be explained. The example is Nagel's. He offered a specific example of an explanation of the law "When gases containing water vapor are sufficiently expanded without changing their heat content, the vapor condenses." It has the form

(T) All As are Cs and All Cs are Bs, therefore All As are Bs.

Now "All As are Cs" cannot be more general than "All As are Bs", for then we would have that "All As are As" is a logical truth, and also that it isn't. Clearly impossible. Moreover, if "All Cs are Bs" is more general than "All As are Bs", then "All As are Cs" is logically necessary. But that is one of the two premises of the explanation, and Nagel's view is that scientific laws are not logically necessary (52–56). The conclusion is therefore that there are no explanations of laws that have the form of (T) – not even the nice one that he provided.

The requirement that for any explanation, what explains should be more general than what is explained still has to be articulated. It might be separated into two parts: (1) those cases when one law is more general than another (even when they may be from two different sciences), and (2) those cases when a theory is more general than what is explained. The explanatory case is our present concern, so the laws case can be set aside. We believe that the problem is still open. We think that there is a partial answer that we discuss in Chap. 12 when the theories that explain are schematic. Then, as we shall note, it seems right to note that any schematic theory is more general than any of the successes that are subsumed under it.[25]

## 8.13 The Model-Theoretic View of Theories – A Disclaimer

The accounts of the explanation of laws that we have discussed so far, have used theories that have been described in more recent literature as linguistically-centered. They are distinguished from the semantic or model-theoretic concepts of theories. Most advocates of this more recent view believe that theories are sets of abstract set-theoretical models, and in the apt phrase of Da Costa and French, proponents of this view wish to "entirely cut models free from Language".[26] The model-theoretic kind of approach has been around now for at least three decades or so. The advocates of that kind of account meant it to be understood as a significant advance over the linguistically-centered theories which consist of some axioms in some formal language - in other words the usual formulations one finds in wide use by scientists to teach, argue and present their results to themselves, their colleagues, and their students. If what the model-theorists claim is correct, then much of what we have to say in the following chapters about theories cannot even be expressed. So it's important to say why we will continue our discussion of theories that are presented as sets of sentences organized axiomatically, but not required to be expressible in some first- or higher-order logical theory.

I should like to say why we have not advocated any of the ingenious versions of the model-theoretic, semantic-based accounts.

There is a rich set of recent papers evaluating the claims of model-theoretic accounts to have distinct advantages over the linguistically based accounts.[27] However we do not find the claims about the advantages of some of the model-theoretic accounts convincing.

---

[25]The topics of explanation and generality are addressed again in sections 12.3, and 12.4, where the notions of explanation using non-schematic theories, and explanation∗ using schematic theories are compared, and a mini theory of the relation of "is at least as general as" is applied to the question of whether theories that explain laws, are at least as general as the laws that are explained.

[26]Newton C. A. Da Costa and Steven French, *Science and Partial Truth, A Unitary Approach to Models and Scientific Reasoning*, p.34, Oxford University Press., 2003..

[27]Cf. H. Halvorson's "The semantic view, if plausible, is syntactic (2013)","Scientific Theories" *Oxford Handbook of Philosophy of Science* (2015), "What Scientific Theories Could Not Be" Philosophy of Science, 79 (April 2012, pp.183–206), and Clark Glymour, "Theoretical Equivalence and the Semantic View of Theories", *Philosophy of Science*, 80,(April 2013), pp.286–297., and J. Azzouni, "A new characterization of scientific theories", published online: 16 May 2014., and Newton C. A. Da Costa and Steven French, *Science and Partial Truth,* Oxford University Press, 2003.

There is the argument cited by Da Costa and French (*Science and Partial Truth*, p.34), that if a theory is just a set of models, that are set-theoretical structures, then no theory can be an object of belief, or have a truth value (true or false) because set-theoretical structures, the models, are not objects of belief and are neither true nor false.[28] They conclude that theories that are identified with a set of models, are neither true nor false, and are not objects of belief. This is a devastating conclusion.[29]

But much worse also follows: On the model-theoretic account of theories, no theory has any logical consequences. None at all. The proof is relatively straightforward:

Suppose that any theory is just a set of models {M1, M2, M3, . . .}, where the models are abstract sets -i.e.: ordered n-tuples consisting of a non-empty set S, together with some relations R1, R2, . . . of various adicities. That is,

T = {<S11, R11, R12, ... >, <S21, R22, R23, ... >, . . . <Sn1, Rn1, Rn2, ... > . . . . }.

The models are sets, and not linguistic items. That is presumably what some advocates of the model-theoretical account adopted, in their wish to avoid representing theories as axioms in some formalized language. That we note is not the account of French and Da Costa.

Now in order for the set T to have some logical consequence C, there has to be some implication relation $\Rightarrow$, such that some (presumably finite) number of the models in the set T imply C. Think of the case where some set of sentences of the set S = {A1, A2, ...} implies some sentence C just in case some of the Ai s imply C. The problem is that there is no implication relation defined for the models. That is, there is no way that some of the models logically imply C. This conclusion is in obvious conflict with scientific practice. Theories are valued for their consequences. Theories that don't have *any* consequences are a non-starter.

In the remainder of this study, we will continue to depart significantly from many theses of the older received view. The first of the departures was already discussed in Chap. 6. That concerned our account of the difference between laws and accidental generalizations,(**LAG**), by making the difference depend on the intertwining of those concepts with that of explanation. Our next few chapters are devoted to theories, schematic and non-schematic, and the special kind of modalities that theories and laws provide.

Chapter 9, our next Chapter, is devoted to David Hilbert's Architectural Structuralism. With David HIlbert's account of theories, we have the first explicit claim that when mathematical or physical theories are properly axiomatized, they are *schematic theories* (our term, not his). They are neither true nor false.[30]

[28]119 The argument is due to Otavio Bueno, conveyed in conversation with Da Costa and French.

[29]Only perhaps if one thinks that all theories have a truth-value. We will challenge that view in Chaps. 9, and 12.

[30]This is not a form of instrumentalism that takes theories to be rules, and therefore lacking in truth value. The items in the schematic theories are sentential.

This important idea has gone almost entirely unnoticed. The explanation for the neglect is that, in an exchange of letters with Gottlob Frege, a key passage was badly translated in a widely known English translation of their correspondence. The impact of this view can be enormous. If it turns out that some of our best theories are not true, then one key starting point for the development of realism about theories is no longer available. Then of course, there is also the problem that if explanations are factive, then the implication seems to be that schematic theories cannot explain, or be explained.

# Chapter 9
# David Hilbert's Architecture of Theories and Schematic Structuralism

*These considerations induce us to conceive of an axiom system as a logical mold ('Leerform") of possible sciences. A concrete interpretation is given when designata have been exhibited for the names of the basic concepts, on the basis of which the axioms become true propositions.*
Hermann Weyl (*Philosophy of Mathematics and Natural Science*, Princeton University Press, 1949, p. 25).

*Thus the axiom system itself does not express something factual; rather, it presents only a possible form of a system of connections that must be investigated mathematically according to its internal [innere] properties.*
Paul Bernays ("Hilbert's Significance for the Philosophy of Mathematics"(1922) tr. By P. Mancosu, *From Brouwer to Hilbert*, Oxford University Press, 1998, p. 192).

## 9.1 The Architecture of Theories

It might be folly to insist on one overall official characterization of the way scientific and mathematical theories ought to be presented. They are usually presented of course in various ways so that the authors and scholars in a field can find optimal means for communication with others, whether within or outside that field. Nevertheless, if theories were thought of as representations of our knowledge, then one way of looking at the representation of theories is to ask, as Hermann Weyl did (2009) "What is the ultimate purpose of forming theories?", and he then cited the familiar proposal of Heinrich Hertz (1894):

> We form images or symbols of the external objects; the manner in which we form them is such that the logically necessary (denknotwendigen) consequences of the images are invariably the images of materially necessary (naturnotwendigen) consequences of the corresponding objects.

This powerful and compact statement of Hertz, resonated with David Hilbert's revolutionary proposal for the axiomatic representation of theories. The kind of



A. Koslow, *Laws and Explanations: Theories and Modal Possibilities*, Synthese Library 410, https://doi.org/10.1007/978-3-030-18846-7_9

axiomatization Hilbert advocated was exemplified in his *Foundations of Geometry*, [1899], and was to become the model for his inquiries into the physical sciences as well as mathematical ones.

Hertz's proposal focused on physical theories and in so far as it considered images (Bilder) of external objects, it would seem to exclude most mathematical theories. In contrast Hilbert's first example of his kind of axiomatization concerned not only the objects and relations of geometry, he maintained that it was also the proper way to represent physical theories as well.

Although Hilbert regarded his *Foundations of Geometry* as a good example of an axiomatized theory, in his sense, his famous description of what he meant by a theory in general does not occur in that work, but in a correspondence with Frege, that is worth quoting a relevant part at some length, for there is much contained in that trenchant passage.

> "... you say that my concepts, e.g.'point', "between',are not univocally fixed. ... But it is surely obvious that every theory is only a scaffolding (schema) of concepts together with their necessary connections, and that the basic elements can be thought of in any way one likes. E.g. instead of points, think of a system of love, law, chimney sweep ... which satisfies all the axioms; then Pythagoras' theorem also applies to these things. All statements of electrostatics hold of course for any other system of things which is substituted for quantity of electricity ..., provided the requisite axioms are satisfied. ...

The first important thing to note is that although the English translation says that a theory is only a "scaffolding (schema) of concepts":

> Sie sagen meine Begriffe z.B. "Punkt", "zwischen" seien nicht eindeutig festgelegt ... Ja, es ist selbstverständlich eine jede Theorie nur eine Fachwerk oder Schema von Begriffen nebst ihren nothwendigen Beziehungen zu einander, und die Grundelemente können in beliebigen Weise gedacht werden.

The original German says that it is obvious that a theory is only a "Fachwork oder Schema von Begriffen". The translation using "scaffolding" doesn't adequately convey Hilbert's meaning. However the Correspondence with Frege was not the first time he described every theory as a "Fachwerk von Begriffen", and certainly not the last. The description occurs in a paper as late as 1918, and the use of "Fachwerk" runs throughout his prepared notes on his seminar lectures, and the lectures which his students prepared. Hilbert's appeal to Fachwerk was frequent and central.

In his written lectures on the Foundations of Geometry (1892), Hilbert described how a theory, a "Fachwerk von Begriffe" can be more or less complete. The sense of completeness meant is not a metamathematical notion. It is connected to the claim that various sciences are associated with special groups of facts. Each science, he thought, aims to order its group of facts with the aid of specific concepts (related to each other he says by the laws of logic). The more complete the group of facts, the more complete the theory. The theory of geometry he believes is complete; it has (now) progressed so far, he thinks, that all of its facts can be derived from a group of earlier ones; there are no new geometrical facts he says. On the other hand, the theories of electricity or optics he thinks are incomplete; there are new facts to be discovered. This notion of completeness has therefore to do with the extent to which theories of a science completely represent the group of facts of that science. This view, sketchy as it is, seems to reflect the influence of H. Hertz's famous remark, already cited, on the way in which a theory represents. Here we shall be more concerned with the representation of theories, rather than the manner in which they represent.

There is in these 1894 lectures a significant addition to what Hilbert wrote to Frege in his 1899 letter. In the letter he said

> But it is surely obvious that every theory is only a scaffolding (schema) of concepts together with their necessary connections, and that the basic elements can be thought of in any way one likes. E.g., instead of points, a system of love, law, chimney sweep ... which satisfies all the axioms; then Pythagoras' theorem also applies to those things.

However in the 1894 Lectures on Geometry he says something similar, but with a helpful addition:

> Ueberhaupt, muss man sagen: Unsere Theorie liefert nur das Scheme der Begriffe, die durch die unabänderlichen Gesetze der Logik mit einander verknüpft sind. Es bleibt dem menschlichen Verstande überlassen, wie er dieses Schema auf der Erscheinung anwendet, wie er mit Stoff anfüllt. Dies kann in der mannigfaltigsten Art geschehen.

Here it is clear that Schema" is used rather than the usual "Fachwerk.". This passage together with the other passages in which "Fachwerk" and "Schema" are used, suggest that any explanation of Hilbert's use of "Fachwerk" should also support the use of "Schema". When he said that it remains for human understanding to apply the schema or Fachwerk to appearences, he says, in this passage, that we ***fill in*** the theory with stuff (mit Stoff anfüllt), with the understanding that the infill can be provided, and can be filled in, in manifold ways.

Hilbert provided various applications of the Fachwerk or schema of geometry. Not only did he furnish examples from every day circumstances (bodies), there were also many examples that made use of algebraic number fields to provide independence results in geometry. There's no shortage of specific applications, and of course there are those famous examples that he suggested which gave some drama to the types of applications he thought possible – as he said in a passage of one letter to Frege cited above

> E.g. instead of points, think of a system of love, law, chimney sweep ... which satisfies all the Axioms; then Pythagoras' theorem also applies to these things. Any theory can always be applied to infinitely many systems of elements.

What is interesting in this particular passage is the hint we have of the way that Hilbert thought of these various applications of geometric theory: they are regarded by him as so many different ways of filling in the Fachwerk or schema. The question worth pursuing is this: what light does this notion of "filling the theory with stuff" shed on his idea that an axiomatized theory is a Fachwerk von Begriffen. Even if all we end with is a more explicit metaphor, that seems to me to still be worth having.[1]

[1]We should note that some translations other than "scaffolding" are a great improvement, though the strict meaning of "Fachverk" is not used. W. Ewald, in his translation (1996) of Hilbert's Axiomatic Thought (1918) translated the phrase "Fachwerk von Begriffen" as "framework of

"Framework" is a significant improvement over "scaffolding" since scaffolds are usually a help in construction. They are usually discarded, however, and not part of the construction itself. It would be bizarre if theories were helpful in "theoretical" constructions, but discarded once those constructions were completed. "Framework" is, we think, much preferred, but there is still the problem of saying what that means.

It is at this point I think we need to take closer look at the possibilities which "Fachwerk" itself has to offer. It's common meaning is that of a timber-wood construction used in homes and other buildings in England, Germany, northern France and Switzerland from the middle ages to the present. In many ways it looks like the framing one might find used today, which is then covered with sheet rock and plywood and then perhaps plastered over. The frame of a Fachwerk however is not supposed to be covered over, nor is it discarded like scaffolding. Rather than concealed, it is supposed to be visible both from the outside and the inside of the construction. The squares, triangles and other spaces or places in the frame can be filled in with stucco, wood, brick, or adobe. It doesn't matter how the places are filled in, so long as the infill conforms to the constraints fixed by the frame.

Before we expand on this picture we should note that there is a certain ambiguity in the claim that a theory is a "Fachwerk von Begriffen". It can mean that once there are specific concepts in place, then a theory is a framework of those specific concepts. Or it can mean that a theory is a framework, and the theory is just the framework, and the infill, the concepts, can be arbitrary –as long as they satisfy the systematic constraints imposed by the framework. It is in the latter sense that we understand Hilbert's view of a theory. The conceptual infill placed within the

concepts, and Mayer and Tilman also use "framework" (2009) in their discussion of Hilbert's Lectures on the Foundations of Physics. David Hilbert and the Axiomatization of Physics: From Grundlagen der Geometrie to Grundlagen der Physik, Springer, 2004. Also we should mention Corey's use of "network of concepts" in his discussion (2004) of Hilbert's axiomatization of Physics.

Fachwerk is subject to the constraints provided by the Fachwerk, but despite the various replacements, the Fachwerk, or theory is supposed to remain the same, invariant, despite the various transformations.

I now want to provide a more obvious logical face to this kind of structure by developing a description of Hilbert's Fachwerk von Begriffen as a schematic axiomatization. There have been several different accounts that have been offered of what it is to be a theory, Hilbert-style, which try to provide them with a contemporary logical cast. Frege thought that Hilbert's axiomatizations were second order concepts – concepts of concepts roughly speaking. That can't be right since concepts for Hilbert aren't individuated by the objects that fall under them. Hilbert thinks of the concepts like point, line and plane as characterized by their relation to each other and to other geometric concepts, and explains how it could be that even pairs of real numbers, or beer mugs could be considered points.

Bernays is more on target, when he says that the axiomatization of Euclidean Geometry is close to the characterization of a mathematical group. He also characterized Hilbertian axiomatizations as being about relational structures and not having a specific subject matter , when he wrote:

> A main feature of Hilbert's axiomatization of geometry is that the axiomatic method is presented and practiced in the spirit of the abstract conception of mathematics that arose at the end of the nineteenth century and which has generally been adopted in modern mathematics. It consists in abstracting from the intuitive meaning of the terms ... and in understanding the assertions (theorems) of the axiomatized theory in a hypothetical sense, that is, as holding true for any interpretation ... or which the axioms are satisfied. Thus, an axiom system is regarded not as a system of statements about a subject matter but as a system of conditions for what might be called a relational structure. (1922).

And, more recently, Stewart Shapiro [2005] has described this as a structuralist kind of position that Hilbert"s work supports. W. Demoupolous on the other hand thought that "...Hilbert, at least in his practice, was quite close to the model-theoretic point of view". A similar view has been expressed by M. Hallett and U. Majer in their introduction to their edition of Hilbert's Lectures on the Foundations of Geometry (2004) in which they refer to the modeling and remodeling that was to become, as they say, "a central part of Hilbert's mature work on geometry, and which often represents a severe distortion of the intuitive meaning of terms like 'straight line'." These views, perhaps with the exception of Frege's, have a measure of plausibility to them. From our present vantage, it does look as if Hilbert anticipated views that were yet to develop, and that these later views could be reflected back, to give a logically detailed description of Hilbert's axiomatizations.

Nevertheless I think there is another way of describing Hilbert's view of theories, that is also contemporary, but closer to his view. It doesn't appeal to the notion of models, or of interpreting and reinterpreting. Instead we want to describe the notion of a Hilbert-style theory as a *schematic theory*.

Suppose we revert to the Fachwerk, with the aim of developing a more abstract or general notion than the literal timber-frame construction. To each of the finite number of places that are framed by the beams of the construction we might give a designated name. To the n of them then, we have $P_1$, $P_2$, . . ., $P_n$. These are the designated places of the construction, ordered in such a way so as to reflect their configuration in the framework. A contractor, in applying this schema, would keep the same framework and fill in the places indicated with various material or stuff.

The range of variation permitted by the theory or its axioms could be given dramatic expression. It is reported by Otto Blumenthal that Hilbert, after hearing a lecture of geometry that was abstract (1891?), on the way back to Koenigsberg, said

> One must be able to say at all times – instead of points, straight lines, and planes –tables, chairs, and beer mugs.

You would think that with such possible variation, the connection with facts in the world might be somewhat attenuated. This wasn't how Hilbert thought of the matter. For him the possible variations were an indication of the variety of applications of the theory, the Fachwerk, and a mark of its fertility.

Less dramatically, Hilbert does indicate that a theory or Fachwerk does have an important connection with the facts (Tatsachen) of a field of science (Wissengebietes). In terms that resonate with H.Hertz's view of theories as a representation of facts, Hilbert says:

> Das Fachwerk der Begriffe ist nichts anderes als die Theorie des wissenschafgebietes: Diese ist ein abstractes Abild, eine Projection der Tatsachen in das Reich der Begriffe[2]

The reference to projections suggests that for Hilbert, a theory is an abstract picture (Anbild) that is a *projection* of the facts of the scientific field into the realm of concepts. And it seems as if the projection he mentions is supposed to be a mapping from the phenomena (Tatsache) of the scientific field to the realm of the conceptual, the theory. If we add to this thought, the idea of theories as representations or pictures of the world, in this case the scientific domain, we arrive at the proposal that what follows factually from other facts is reflected by the projection so that the consequences of the projections (pictures or images) of the facts, the appropriate parts of the theory, is invariably the projection (picture or image) of naturally necessary consequences of certain facts from others. The following diagram suggests a close alliance of Hilbert's thought about how a theory is related to a scientific field with Hertz's idea of how a theory represents the world:

$$\begin{array}{llll} \underline{\text{Theory}}\,(\text{Fachwerk}): & \varphi(\mathrm{f}) \Rightarrow \varphi(\mathrm{g}) & (\mathrm{s}) & [\text{logically necessary consequence}] \\ & \varphi\uparrow \qquad \varphi\uparrow & & \\ \underline{\text{Scientific field}}: & \mathrm{f} \quad \rightarrow \quad \mathrm{g} & & [\text{naturally necessary consequence}] \end{array}$$

Here we let φ be the projection which maps a fact f of the scientific field to its image or picture, φ(f), in the theory of that field. According to Hertz (1894, p.1), any logically necessary consequence (denknotwendigen), say s, of φ(f) will be a projection of some fact g which is some naturally necessary consequence (naturnotwendigen) of the fact f. And conversely, if g is a naturally necessary

[2] *Hilbert's Lectures on the Foundations of Physics*, 1915–1927,bT. Sauer and U. Majer (eds.), Springer 2009, p. 420.

consequence of some fact f, then its projection in the theory will be a logical consequence of the projection of f.

We have suggested that this account of the way in which theories represent a scientific field is obviously very close to Hilbert's view of representation, – once one draws the parallel between the images or pictures of objects of Hertz on the one hand, and the projections of facts of Hilbert on the other.

However we should note that this suggestion is just that. The evidence is slender – just the opening statement on the first page of Hertz's Principles of Mechanics, and the later similar remark in section 428 of the same work in which he stated that

> The relation of a dynamical model to the system of which it is regarded as the model, is precisely the same as the relation of the images which our mind forms of things to the things themselves. For if we regard the condition of the model as the representation of the condition of the system, then the consequents of this representation, which according to the laws of this representation must appear, are also the representation of the consequents which must proceed from the original object according to the laws of this original object. The agreement between mind and nature may therefore be likened to the agreement between two systems which are models of one another, and we can even account for this agreement by assuming that the mind is capable of making actual dynamical models of things, and of working with them,

together with the brief and cryptic statement of Hilbert's that we cited, according to which a Fachwork der Begriffe is an abstract picture, a projection of facts into the realm of concepts.

Nevertheless, the mathematical beauty of the idea is striking, and its scope is wide, given that Hilbert intended the idea to cover both mathematical fields as well as those of physics and even biology. Hilbert also thought that the mapping or projection was one-to-one, and that would make the mapping into an isomorphism of sorts. Aside from interesting questions about the mapping, the logical necessity has since then been studied in depth. The other half of the representation concerns natural necessity (naturnotwendigen), and that notion of necessity is still an open problem for philosophers of science.

Interesting as the pursuit of this kind of representation is, we want to focus here not on the way in which axiomatizations represent a scientific field, but the character of the axiomatizations themselves. Consider a typical example of some axioms that are close to those to those given in Hilbert's axiomatization of Euclidean Geometry but taken from a later version by Hilbert and Bernays[3] This axiomatization unlike Hilbert's does not use the concepts of points, lines, and planes, but uses only just that of points, and two relations: Gr(x, y, z) for the relation "x, y, and z lie on a straight line", and Zw(x, y, z) for the relation "x lies between y and z". The English readings below are not part of the axiomatization.

(1) $(x)(y)(z)\,(Zw(x, y, z) \rightarrow Gr(x, y, z))$. If x lies between y and z, then x, y, and z lie on a straight line.

[3]D. Hilbert and P. Bernays, *Grunndlagen Der Mathematik*, vol 1, verlag von Julius Springer, 1934, pp. 5–7.

(2) (x)(y) ¬Zw(x, y, y). No x lies between any y and y.
(3) (x)(y) (x $\neq$ y $\rightarrow$ (Ez)(x, y, z) ). If x and y are distinct points, then there is a z such that x lies between y and z.

These examples are not the full set of axioms concerning "between", but they give some sense, of what Hilbert meant when he observed that the axioms for betweeness also hold for the electrochemical series of metals, and for the laws of heredity. As Hilbert noted (1930) in a discussion of the variety of applications of the axiomatic method, consider the breeding of drosophila:

> But now with suitable cross-breeding later generations exhibit a smaller number of deviations from the commonly occurring couplings – indeed, the percentage is a definite constant. The numbers which one thus finds experimentally tally with the linear Euclidean axioms of congruence and with the axioms for the geometrical concept 'between', and so the laws of heredity result as an application of the axioms of linear congruence, that is, of the marking-off of intervals; so simple and exact, and at the same time so wonderful, that probably no fantasy, no matter how bold, could have devised it.

How then does some bit of hereditary theory become an application or special case of part of the axiomatization of Euclidean Geometry? We have suggested earlier that in a Fachwerk, the places $P_1$, $P_2$, ..., $P_n$ in it could be indicated by place names which would enable one to systematically describe where the infill was to be placed.

Analogously, since any axiomatization is thought of as a Fachwork of concepts, there will be those places in the axioms in which the concepts of the theory are "positioned" or placed with respect to each other. Those positions are but are place holders in the theory, and we shall mark their occurrences by the presence of letters $S_1$ , ..., $S_m$. These we shall call these schematic predicate letters (schematic letters for short). The schematic letters are to be replaced by specific predicates and relations in the axioms. I leave to one side a number of qualifications that need to be made –eg. the right kind of adicity for the replacements.[4]

There is an apparent problem that should be immediately addressed. Hilbert thought that his axioms did not have a truth value, and believed that concepts such as "point", "line", "plane", and "between" are not fixed by the objects that fall under or instantiate them, but they are instead individuated by their relations to each other.

It is fairly obvious that inspection of typical axioms that Hilbert and Bernays provided for Euclidean Geometry on the face of it, have no schematic letters in them. Thus, for example, in

(1) (x)(y)(z) (Zw(x, y, z) $\rightarrow$ Gr(x, y, z))

there is no gap. There is no obvious position which can be "filled",and certainly no obvious schematic letter. "Zw" and "Gr" seem to be specific relations ("between" and "collinear") rather than uninterpreted expressions. But they must in some sense

[4]The use of schematic letters rather than variables of the appropriate type, allows them to be replaced by specific predicates and relations. This would not be permitted if the places in the Fachwork were marked by variables of the appropriate type.

be uninterpreted in order to accommodate Hilbert and Bernays' view that (1), and the entire axiomatzation of which it is a part, lack truth value.

There are at least three ways of meeting the condition, but I think that only one of them is sufficiently close to making good on Hilbert and Bernays' thinking of axiomatizations as Fachwork or Schema. The first suggestion is to take "Zw" and "Gr" as specific relations (with definite extensions) and to interpret them in the sense that there is a function that maps predicates and relations to predicates and relations, so that Zw* and Gr* are assigned to Zw and Gr respectively, and as a result (1) gets mapped to

(1)* $(x)(y)(z)\ (Zw^{*}(x, y, z) \rightarrow Gr^{*}(x, y, z))$.

In this case the mapping carries predicates (relations) to predicates (relations), and the result, on replacing the predicate by it's value under the mapping is (1)*.

The second suggestion is just to deny that "Zw" and "Gr" are specific relations. They are meant to be schematic letters, and the different specific relations "Zw*" and "Gr*" just replace them, without invoking the mapping that the first suggestion used.

The third suggestion is the one that I think is the best fit. It says that one should think of the axioms as having different schematic letters or place holders and the idea is that one replaces them systematically throughout the axioms by the specific relations "Zw*"and "Gr*". It is the third suggestion which thinks of the schematic letters as place holders in axioms for the various possible infill. The axioms with their place holders do not have a truth value. However, when specific replacements are made, the results are *applications*, or cases of examples of the theory, and each of them has a truth value.

We should note that we follow Bernays (1934) in his later discussion in Hilbert and Bernays, where he says of these basic relations "lies on" and "lies between" are not names in the axiomatization. The use of these terms is, he says, a concession to intuitive geometry. We do not follow him however when he says that in formal axiomatics they should be regarded as variable predicates (sic) [Hilbert and Bernays, 1934]. Nevertheless, we do think that Hilbert's stress on the notion of a Fachwerk in which specific predicates fill in the places in the Fachwerk is closer to our use of "schema". And of course, "Schema" is the term that Hilbert himself employed.

Thus we end up with a theory or Fachwerk as a kind of theoretic schema. This has the added virtue of not having to worry whether there is some difference in Hilbert's use of both "Fachwerk" and "Schema". A Fachwerk is a kind of schema.

## 9.2 Hilbertian Concepts

We have argued that theories á la Hilbert, are schematic. That makes it obvious that they have no truth value.

This raises another problem that has not received much attention. What can one make of his concepts? It is clear from his correspondence with Frege that he doesn't distinguish concepts by the objects that fall under them. He affirms instead that they are distinguished by their relation (in the theory) to the other concepts of that theory. And in fact he thought that the concepts were *defined* by the axioms of the theory, and he asserted, as the following passage from his correspondence with Frege makes evident, that every new axiom to a theory changes the concept.

> Therefore: the definition of the concept point is not complete till the structure of the system of axioms is complete. For every axiom contributes something to the definition, and hence every new axiom changes the concept. A point in Euclidean, non-Euclidean, Archimedean and non-Archimedean geometry is something different in each case.

It is widely thought that Hilbert meant that the axioms of a theory *implicitly defined* the concepts of that theory. Bernays thought that, although he noted that Hilbert never used the German equivalent of that term. And in any of the senses that term has for us now, it does not seem that this notion will do. Hilbert (in correspondence with Frege), explained that it should be possible for

> ...the basic elements [of a theory], (added the author) instead of points, think of a system of love, law, chimney sweep ... which satisfies all the axioms; then Pythagoras' theorem also applies to these things.

It appears then that Hilbert thought that there were ways in which the predicates and relations in the theory could be replaced by very different ones that would still satisfy the theory. However that is not possible on familiar accounts of implicit definition.

For example, on one account of implicit definitions, if we have a theory $\Omega[P(x)]$ in which the predicate "P(x)" occurs, then the theory is said to implicitly define that term, if and only if, for any predicate "P*(x)" uniformly replacing all the occurrences of "P(x)" in $\Omega[P(x)]$,

$\Omega[P(x)] \wedge \Omega[P^{*}(x)]$ logically implies $(x)(P(x \equiv P^{*}(x)]$.

That is, the conjunction of the two theories "P(X)" and "P*(x)" would have to be coextensional –and clearly the examples that Hilbert provided show that this is not so. If the axioms do not always implicitly define the predicates in them, then how else might they be related?

The first thing we want to say is that it is probably better to focus on explaining the notion of something's being a *concept of a theory*, rather than the notion of a concept *simpliciter*, without mentioning it's theoretical context (if it has any). I don't think Hilbert had much to say about concepts in general. We shall focus instead on "C is a concept of theory T" rather than "C is a concept. We shall simplify things a bit by considering a single axiom as our theory, say

(1) $(x)(y)(z)\ (Zw(x, y, z) \rightarrow Gr(x, y, z))$,

following Bernays' account of Euclidean Geometry in Hilbert and Bernays.[5] When we describe the structure of this axiom, we use the terms "between" and "collateral", so that (1) is read as "If x lies between y and z, then x, y, and z are collinear. Let us write (1) as $\Omega_1[Zw, Gr]$. For any predicate $R(x, y, z)$ we shall say that **R is a**

[5]Hilbert and Bernays, vol.1, p.6, Axiom II,1.

**between-concept of $\Omega_1$** if and only if for some predicate S(x, y, z), ("S" for short"), (x)(y)(z) (R(x, y, z) $\rightarrow$ S(x, y, z) ). Thus our proposal is that

$$\text{R is a between - concept of } \Omega_1 \text{ if and only if } \Omega_1[\text{R, S}], \text{ for some S.}$$

It is clear that for different axioms of Euclidean geometry we need only change $\Omega_1$. We have not referred to a between concept (*simpliciter*), but to the between-concept of a particular axiom, and the between concepts of different axioms will in general be different. If we take all the axioms into consideration, say Ω[Zw, Gr] (Hilbert and Bernays 1934), formulated with the use of only the betweeness and collinearity relations, then we can drop the reference to particular axioms and define "R is a between-concept of Ω" and "R is a collinearity-concept of Ω as follows":

(i) R is a between-concept of Ω if and only if Ω[R, S], for some predicate S, and
(ii) R is a collinearity-concept of Ω if and only if Ω[T, R], for some predicate T,

where it is understood that R, S and T, are all three-place predicates. (i) and (ii) are necessary and sufficient conditions for any three-place predicate to be a between-concept or a collinearity-concept of Ω.

This suggestion does not show that the axioms define any particular between-concept or collinearity-concept of the axiomatization. We don't have an explicit or implicit definition. What we do have is something close, but different. We have definitions of the property of being a between-concept and the property of being a collinearity-concept in terms of the axiomatization.

## 9.3 Schematic Physical Theories

Hilbert's view of theories as schematic, holds for all theories, mathematical, physical, biological, and perhaps even more widely. Here is Hilbert on the grand design, as expressed by Bernays:

> Everything whatsoever, that can be the object of scientific thought is subject, as soon as it is ripe for the formation of a theory, to the axiomatic method and thereby of mathematics. [Mancosu, tr., Bernays 1922].[6]

That is the very ambitions program that Hilbert promoted. There is one consequence of it in particular that Hilbert not only acknowledged, but emphasized as a virtue.

In his correspondence with Frege, Hilbert noted that according to his Fachwerk account of theories,

> ... any theory can always be applied to infinitely many systems of basic elements. One needs only to apply a reversibile one-one transformation and lay it down that the axioms shall be correspondingly the same for the transformed things.

[6]This was a longstanding view of Hilbert. Recall that his list of open problems in 1900 included the sixth problem: Axiomatize all of Physics.

> ... the circumstance I mentioned can never be a defect in a theory,* (* It is rather a tremendous advantage.) and it is in any case unavoidable. (41)[7]

This is a truly remarkable aside by Hilbert. Not only does he tell Frege that there is no one system of basic elements of a theory, but on his account there are many, even infinitely many applications with different basic elements. Moreover with the brevity and incisiveness one gets from one mathematician to another, there is a proof. It is exactly the same theorem that Newman appealed to in his critique of Russell's structural realism.[8] It is closely related to the argument that Davidson used to establish referential indeterminacy, and that Putnam used in what has become known as "Putnam's Paradox". What is remarkable here is that early on Hilbert acknowledged it as an unavoidable consequence of his account of theories, and claimed it as a tremendous advantage of his account.

This is a consequence which Hilbert acknowledged and fully endorsed. There are other features however, which he did not mention, but which may be more troubling:

(i) If scientific theories have any use at all, they are surely supposed to offer explanations. However, explanations are usually assumed to be factive. Any statement that is part of an explanation has to be true. If we let Exp[A; B] stand for the existential claim that A is all or part of some explanation of B, then facticity, it will be recalled[9], requires that

$$\text{Exp}[A; B] \Rightarrow \quad A, \text{ and } \text{Exp}[A; B] \Rightarrow B.$$

So, theories, on the Hilbertian account of them, do not explain anything, nor can they be explained.

(ii) It is widely thought that theories count among their successes, those applications which supposedly follow deductively from them- eg. Newton's deductions of Galileo's law of falling bodies, Kepler's laws of planetary motion, the laws of pendulum motion, the two tides on the earth's ocean, and much more. These would all count as successes of the theory. However, if we thought that Newtonian classical mechanics was best described as a schematic theory, something which we believe, and will explain why, in greater detail in the following chapter, then it seems that none of these famous applications of that theory follow deductively from it. That is because none of the instances of a schema follow deductively from it (modulo of course those instances of the schema that are logical truths).

These features of schematic theories may seem like very strange business, and in consequence, the requirement that the representation of our theories be schematic,

---

[7] *Philosophical and Mathematical Corresspondence: Gottlob Frege*, The university of Chicago Press, Chicago, 1980. Eds G. Ga briel, H. Hermes, F. Kambartel, c. Thiel, AS.Veraart. Abridged from the German edition by B. McGuinnness and translated by H. Kartel.

[87] Newman, M. H. A. (1928), "Mr. Russell's 'Causal Theory of Perception,'" *Mind*,37(146): 137–148.

[9] Chap. 7.

might not have much credibility in light of current practices in the sciences. There's no doubt of the presence and power that schematic theories have in logic and in mathematics. Hilbert's claim that *all* scientific investigations *that are ripe for a theory* should be represented schematically seems, on the fact of it, ill-suited advice.

Nevertheless, I want to suggest that this adverse conclusion is not warranted. There is much to be said for going schematic, at least for some theories. Moreover, as we shall see, some very important physical theories are already schematic.

I want to suggest some considerations that address what appear to be several troubling consequences.

The first consideration is that many successful theories are in fact naturally thought of as schematic so it's not a matter of forcing them into some procrustean Hilbertian mold; they are already best formulated as schematic. This situation we shall consider in the following chapters.

The second is that since schematic theories do not have a truth value, they do not figure in any explanations. I will argue that that nevertheless they still have explanatory power.

The situation in which the successes of a theory are nevertheless not deductive consequences of it, is not something peculiar to schematic theories. It is also a consequence of an earlier view, – one which I think is mistaken, that theories are best represented by their Ramsey sentences.

Basically the idea is to take a preliminary formulation of any theory T that is expressed with first order quantifiers, various predicates and relations (some "observational" and some "theoretical") and the usual logical particles. Let that sentence be given by $\Omega(P, Q, R)$, where we have suppressed any mention of the observational predicates and relations, and list only the theoretical ones. The Ramsey sentence of the theory, R(T), is the result of replacing the theoretical predicates and relations by the second order variables "X", "Y" and "Z", and prefixing the second order existential quantifiers with respect to those variables. The result then is $(\exists X)(\exists Y)(\exists Z)\Omega(X, Y,Z)$. If we count as the successes of the Ramsey sentence those cases $\Omega(U, V, W)$ for the specific predicates or relations U, V, and W, which make the Ramsey sentence true then it follows that none of them are deductive consequences of the Ramsified theory, for then the Ramsey sentence would be equivalent to $\Omega(U, V, W)$, since then R(T) implies $\Omega(U, V, W)$ and $\Omega(U, V, W)$ also implies R(T)). And it also is apparent that the Ramsey sentence doesn't figure in any deductive explanation of them. However for those who still think that Ramsey sentences are a viable way of representing physical theories, one can point out that they share some features with the schematic proposal: Not every sentence containing theoretical terms will have a truth-value – though the reason for this rests on Ramsey's way of attributing truth to theoretical statements, and not on schematic considerations. On the schematic view, those terms are regarded as schematic predicate letters. On the Ramseyan view, that sentence will only have a truth value if the result of placing it within the scope of the existential second order quantifiers, results in a true Ramsey sentence. If that theoretical sentence is outside the scope of the quantifiers, it has no truth-value. Within the scope however, it can turn out that

that sentence will have both the value "true", and also the value "false" (as pointed out by van McGee[10] and later by A. Koslow[11]. A third observation worth making is that there are some powerful theories, apparently non-schematic with many successful instances to their credit, for which it turns out that those theories do not figure in the deductive explanation of any of their successes (nor do they figure in any probabilistic explanation of them for that matter). Some examples are The Germ theory of Disease, the Kinetic theory of gases, and Hamilton's Principle of least action. They were singled out in S.Morgenbesser's trenchant description of theories that are *theories – for* the systematization of their instances, but they are not *theories – of* any of those instances.[12]

## 9.4 Schematic Theories, Schematic Subsumtion, Subsumptive Explanation and Successes

Ramsey-style theories, and *theories -for,* like the Germ Theory of disease, and the Kinetic Theory of Gases, are unlike the schematic theories in having truth values. But they share with schematic theories the fact that they have successes that are not deductively (or probabilistically) explained by them. That tells us that aside from the issue of truth-value (if it is one), the other features of schematic theories are not peculiar to them alone.

A few closing observations: (1) How bad is it that schematic theories don't have a truth value? I don't think this is a calamity. After all one can still study their deductive properties, and carry out complex deductive theoretical arguments with them. Model theoretic studies of them require some subtlety but they too are still available. Moreover, the result of specific replacements of the schematic letters, their applications, do have truth values. So the matter of whether theories are true or false is a subtle issue. Schematic theories have no truth-value, but their applications do. Schematic theories yield a nice split decision.

(2) How bad is it that schematic theories do not figure in explanations of their applications. The special case of Newton's theory of mechanics is, we believe, a schematic theory, where the "F" in the second law of motion is construed as a schematic function term. The idea is that "F" in the expression of the Second Law of Motion is not a name of one special force function – in that case, the second law would be about one special case. And it is not a function variable (for essentially technical reasons having to do with the understanding of variables. The "F" is, we

[10]"Ramsey's Dialetheism", in *The Law of Non-Contradiction,* G. Priest, B. Armour-Garb, and J.C. Beall (eds.), Oxford University Press, 2005.

[11]A. Koslow, "The Representational Inadequacy of Ramsey Sentences.", *Theoria*, vol LXX!II, 2006, Part 2.

[12]S. Morgenbesser and A. Koslow, "Theories and their Worth", *Journal of Philosophy* 2012, pp.616–647.

believe, a schematic function letter which in various cases, for specific mechanical systems, it is replaced by particular functions, not by the names of those functions. Newton's theory of Gravitation is a special application of the schematic theory of mechanics, in which "F" in the theory of mechanics is replaced by the usual gravitational force function. For this particular example, the schematic Newtonian theory of mechanics explains nothing, being schematic. However it has an application, the Newtonian theory of Gravitation, which explains the three Keplerian laws of planetary motion, Galileo's law of falling bodies, the theory of the tides and so on. Consequently, although schematic theories, strictly speaking, don't explain. They do have applications which certainly do explain.

Now it does seem strange to credit explanatory successes to the applications of a schematic theory but to deny crediting them to the theory itself. However there is no need to do that. We can see that there is a type of subsumtion to be had, by using a variation of a notion of subsumtion that Hempel once appealed to when he said

> ... I think that all adequate scientific models and their everyday counterparts claim or presuppose at least implicitly the deductive or inductive subsumability of whatever is to be explained under general laws or theoretical principles. [*Aspects of Scientific Explanation* 1965, 424–5].

The subsumtion Hempel had in mind, subsumtion under a theory T, was confined to being either a deductive or probabilistic consequence of T. In the case we have been considering, schematic or place-holder theories, there is another kind of subsumtion to be had, which is neither deductive nor probabilistic: *schematic subsumtion* of the applications of a schematic theory under T.

By "schematic subsumtion" we shall mean a relation between a schematic theory T and a sentence S:

S is subsumed under, T ( T >> S) if and only if
(1) T is a schematic theory, with schematic predicate letters S1, S2, ... , Sn.
(2) T* is the result of replacing the schematic predicate letters S1, S2, ... , Sn, by the specific predicates S1*, S2*, ... , Sn* respectively, and
(3) T* is equivalent to S.

Given (1) – (3), we shall say that S is (schematically) subsumed under the schematic theory T.

Consequently, the various applications that are subsumed under a schematic theory that are equivalent to laws, are also subsumed under that theory. They are all special cases that result from the replacement of the schematic letters in the theory by specific predicates, relations, and functions. They all fall under the theory in a uniform way; only that way is not deductive.[13]

[13]The uniformity involved in the various subsumtive applications is radically different from the deductive uniformity studied by P. Kitcher,"Explanatory Unification", *Philosophy of Science*, 48, 507–31, and M. Friedman, "Explanation and Scientific Understanding", *Journal of Philosophy*, 71(I): 5–19, 1974. For one thing the uniformity that Kitcher appeals to is a uniformity of a deductive pattern, while all the subsumed cases of a theory are not deductive consequences of that theory.

With this notion of schematic subsumtion in place, we shall propose that for any applications that have explanations to their credit, then that credit also extends to the schematic theory under which it is subsumed. That is, in terms of *schematic subsumtion* (or *schematic covering*), the idea of *subsumptive explanation* is this:

**(SS)** Let T be a schematic theory. Then T *subsumtively explains* (hereafter, *explains**) some A if and only if there is some non-schematic sentence S that is schematically subsumed under T, and S explains A.

The point is just the simple one of giving credit where credit is due. Subsumtive explanation isn't fully factive. If a schematic theory T subsumtively explains A, it doesn't follow that T is true. T is schematic. But it does follow that A is true since a non-schematic sentence S explains A. So we have the kind of factivity that "knows that A" enjoys.

Schematic theories can have two kinds of successes. Let T be a schematic theory. Among its successes then there are those As that are explained* by T, and then too there may be some laws L that are subsumed under T i.e. T >> L.

Schematic theories are among the most powerful and prized theories that we have, and their scientific value resides especially in the huge number of successes, – explanations* and laws that are subsumed under them. We will study them in

greater detail in Chap. 12, where it will be shown that if a schematic theory T explains* a contingent generalization E, then it follows that E is not only true, but is also non-accidental. That is, schematic theories have the same virtue in this regard as non-schematic theories.

We introduced the topic of schematic theories in order to explicate E. Nagel's account of theories, by arguing that the best formulation of it required the representation of theories as schematic. In the present chapter we gave a more stand-alone account of schematic theories for two reasons: (1) It represents Hilbert's view that all properly axiomatized theories, mathematical or physical, are schematic. This important aspect of Hilbert's view of theories is – to understate it – not as well known as we think it should be. And, (2) There are many powerful physical theories (e.g. Newtonian Classical Mechanics, Lagrangian Mechanics, Hamiltonian theory, and Probability theory (in Kolmogorov's version)) to mention just a few, that are in fact schematic. We believe that philosophical analysis, and an appeal to scientific practice would make that evident for many other physical theories as well.

In the following two Chaps. 10 and 11,we will show that theories provide a special kind of modal possibilites, and that laws associated with those theories also have a special kind of nomic possibilites associated with them.

One obvious way to show this, would be to appeal to the various states of systems that theories describe, and maintain that those states provide the physical modalities- i.e. the states that are associated with theories are the physical modals that the theories provide and in addition, a subset of those possibilities are the nomic modal possibilities associated with the laws of that theory. The appeal to the states of theories would be a good start for this general discussion if only there were a common concept of state shared by various theories.

Unfortunately, a brief survey of various sample theories and their states, reveals such a multiplicity of different concepts of state, that this prospect looks hopeless. Nevertheless, we think it is possible to make some headway. We will show that for different theories (and theoretical scenarios) it is possible to construct *in a uniform way* for each theory T, a *magnitude vector space* (**MVS**) associated with T. Each magnitude space is a vector space that is constructed in the same way from the various physical magnitudes provided by each theory or theoretical scenario. Finally, for each magnitude space vector space we shall show how to associate an *implication structure* of a theory T – essentially a set based on the magnitude vector space of the theory T together with an implication relation on it.

We will then show that for any theory T, there are defined physical possibilty *operators* that are functions ***on*** the implication structures associated with the theory T, and there are also modal entities ***in*** the implication structure associated with the theory T. The distinction between the modal operators on an implication structure, and the modal entities in that structure is novel, and we hope of some further interest. We will also describe the special features of these modals that helps to locate them with respect to a larger group of more familiar modals. We hasten to add however that these modal operators and modal entities are indexed to specific theories, and will in general be different for different theories.

# Chapter 10
# Theories, Theoretical Scenarios, Their Magnitude Vector Spaces and the Modal Physical Possibilities they Provide

With this chapter we begin the development of a view about those scientific laws each of which has associated with it, a background that consists either of some theory, or a loosely knit collection of statements that involves various physical magnitudes that are involved in the expression of the law. We shall refer to the latter kind of background as the *theoretical scenario* for the law.[1]

As will be made clear in what follows, one role of these theories and theoretical scenarios is to provide a collection of magnitudes that will be used to define the notion of theoretical possibilities, which in turn will play a role in explaining how laws are connected to modal possibilities. Perhaps then, a few examples of what we mean by such scenarios will be helpful.

## 10.1 Theoretical Scenarios and Laws

We think of theoretical scenarios as sometimes being theories, and sometimes something referring to something less than a theory. The connection between laws and their theories or theoretical scenarios is complex. Here are some familiar examples that illustrate that complexity:

(i) **Newton's theory of Mechanics** consists of three laws. The first, the Law of Inertia, the second, roughly stated, that the (external) force on a body is equal to

[1]In our discussion of the difference between laws and accidental generalizations Chap. 7), it was assumed that there was some explanation of the instances of laws (and in our discussion of Braithwaite's account (Chap. 8) it was assumed by him that laws were always embedded in some deductive system. Here we assume something weaker: a theory or a theoretical scenario for each law, that provides a background of physical magnitudes, but we do not assume that it provides enough information to guarantee an explanation or even a deduction of the law.

 
A. Koslow, *Laws and Explanations: Theories and Modal Possibilities*, Synthese Library 410, https://doi.org/10.1007/978-3-030-18846-7_10

the body's mass multiplied by its acceleration, and the third, that the action of a body is equal and opposite to the reaction.[2]

(ii) **Newton's Law of Gravitation** is another matter. It requires that for any two bodies, the gravitational force between them is the product of their masses divided by the inverse square of the distance between them, and directed along the line connecting their centers. It has Newton's theory of Mechanics (1) as a theoretical scenario. As we shall see below, Newton's Law of Gravitation is not a consequence of his theory of mechanics, so that in this case, we have a law that is not a logical consequence of its theoretical scenario.

We can gather more detail in particular about the relations between laws and their theoretical scenarios when considering the Keplerian laws of planetary motion and Galileo's law of falling bodies prior to Newton's explanation of them:

(iii) **Kepler's three laws of motion.**

Kepler thought that there was a kind of magnetic attraction of the planets toward the sun, and that there were also spokes or special celestial filaments extending from the sun, that rotated with the sun at the hub of these aetherial spokes. It was this combination of the action of the spokes hitting the earth and other planets, combined with the attraction of them towards the sun that produced (somehow) an oval orbit around the sun. Once Kepler had determined the kinematic description of the planetary orbits, the task was to explain why that was so. As Dijksterhuis concisely, but aptly puts it,

> Kepler was not now going to content himself with the description of what takes place in the sky, but he was going to find also the causes by which they are called into existence; he was going to add a dynamic theory of celestial motions to the kinematic theory he had already given. (p.309)[3]

Nowadays we would hardly count this background scenario as an explanation of the Keplerian laws, but it was surely provided by Kepler in order to show why the elliptical orbits which he had already determined was a theoretical possibility –even if that scenario did not provide an explanation. Nevertheless, in trying to provide a dynamical theory for planetary motion, aside from introducing some general but false claims (e.g. the Law of distances),[4]Kepler introduced terms for new magnitudes of his proposed theoretical backing when he claimed that

---

[2]We will present a more exact discussion below.

[3]E.J. Dijksterhuis, The Mechanizaton of the World Picture, Tr. by C. Dikshoorn, Oxford University Press, 1961.

[4]It is a false generalization, which deductively yields Kepler's second law about equal areas being swept out in equal times. It says that if a planet moves on a circle with center C, and r is the radius from the planet to a point E off the center C (the center of an eccentric circle), then the velocity of the planet is proportional to the inverse of r. Kepler seems to have thought that this law holds even when planets move on ellipses.

> If the word soul (*anima*) is replaced by force (*vis*), we have the very principle on which the celestial physics in the Mars-commentaries (i.e. the *Astronomia Nova*) is based.[5]

Evidently there were old magnitudes as well as new ones that figured in the theoretical scenario. Thus, even in the important case of Kepler's laws of planetary motion, there was still a theoretical background, –though not the one provided by Newton later on, which used certain magnitudes that proved to be important in the description of those laws.

(iv) **The Galilean corpus** yields examples of laws that had theoretical scenarios that did not yield explanations. Any discussion has to be very carefully described, since some generalizations, such as the law of squares,[6] was known to Galileo, as early as 1605 (according to his letters),[7] but he also tried at one time to derive it from the false hypothesis that the instantaneous velocity was proportional to the distance traversed. This attempt to back up the law of squares was later supplanted with another attempt using the notion of impressed forces acting on the falling body. Thus we have a generalization being backed up by theories that involved magnitudes that were a mixture from varied sources borrowed from the science of an earlier time. As Dijksterhuis correctly pointed out, the law of squares was later derived in Galileo's Discorsi using the correct relation between the distance covered and the time elapsed (not the space traversed). The derivation relied on the body receiving a series of impetuses at evenly spaced intervals of time. The derivation itself however, was faulty, though the background scenario was flush with a rich supply of physical magnitudes.

These examples are cases where the laws were once part of theories, – only the theories were just incorrect. It turned out that in fairly short time, the Galilean and Keplerian laws were separated from their original background theories and acquired new theoretical scenarios. The laws became orphans for a time - only to acquire new theoretical scenarios later, when Newton's Theory of Gravitation became the theoretical scenario for both the Keplerian and Galilean laws.

It might be thought that the theoretical scenario for a law has to be on the scene prior to the law. This is not generally so. Ohm's law, as we shall see, is a case in point, where the novel theoretical scenario was introduced at the same time with the law.

(v) **Ohm's Law** is a very powerful tool in solving complex problems in circuitry and its scientific merit and usefulness are not in dispute, even though it has limited scope, outside of which, it is unreliable. Georg Ohm introduced the law that $V = R \times I$ in 1827, and according to J.C. Maxwell, simultaneously provided a theoretical scenario of a group of new concepts that he used to express that law:

[5]Dijksterhuis, 310.

[6]Roughly, that the distance traversed by a falling body is proportional to the square of time elapsed.

[7]Dijksterhuis (339).

- "electromotive force" (V for volts nowadays, though Ohm used "E"), "strength of current (I)", and "electrical resistance" (R) replaced the vague, non-quantitative concepts used prior to his paper.[8] Thus I regard it as a case where the theoretical scenario and the law are introduced in one fell swoop.

We now want to define a notion of a state of a theory (or of a theoretical scenario), which will be available for many, if not all theories. It will certainly be available for all those theories or scenarios that specify a non-empty collection of physical magnitudes. It is worth first to have a brief survey of what passes for the states of a theory, when the variety of theories is taken into account.

## 10.2 The Variety of States of Theories

Not that much has been written about the notion of states of systems or the states of theories in general, though a great deal has been written about the states of many particular physical systems and many theories. There are a few scientists and philosophers who have considered the matter and have provided some seminal hints at what a reasonable though rough general account might look like. Here are a few of some seminal observations about the relation between laws and the states of systems and theories.

**(2.1)** Percy Bridgman[9] noted that a body is in a "state" (sic) when its properties have a certain value. This seems to be a rough description of what is common in thermodynamics, but Bridgman was quick to add that (1) "The notion of "state" is prior to the laws of thermodynamics, in fact if it were not prior the first law of thermodynamics, he maintained, would be reduced to a tautology (sic)." He then asked the question on which his general notion of state rests:

> Do bodies have "properties" (sic) such that when they are fixed the behavior of the bodies is fixed.

Bridgman makes no mention of laws or their relevance to the states of a system. The obvious question is "What determines which properties are the "fixers" of behavior.

**(2.2)** James Cushing[10] in considering elementary quantum theory, describes a notion of state, presumably designed to cover many physical theories generally.

> In a physical theory we typically describe a system (sic) in terms of its *state*. We specify the relevant physical quantities or variables and then use dynamical laws to find the time evolution of these variables to predict their values in the future.

---

[8]Cf. the incisive discussion by J.C. Maxwell "On Ohm's Law". Reprinted from the *British Association Report*, 1876, reprinted in *The Scientific Papers of James Clerk Maxwell*, ed. W.D. Niven, MA., FRS. Two volumes bound as one volume. Volume 2, pp. 533–537.

[9]*The Nature of Thermodynamics*, Harvard U. Press 1941, Harper Torchbooks 1961, p.17.

[10]*Philosophical Concepts in Physics*, Cambridge University Press, 1998, p.290.

Cushing, in contrast to Bridgman, emphasized that it is the *laws of a theory* that provide the means of prediction of values of the properties or magnitudes which presumably specify the state at any time. That is a possible role that laws can have.

**(2.3)** A.B. Pippard[11] in a discussion of the second law of thermodynamics made an observation about the first law of thermodynamics that could easily be regarded as appropriate for laws generally:

> The first law of thermodynamics expresses a generalization of the law of conservation of energy to include heat, and thereby imposes a formidable restriction on the changes of state which a system may undergo, only those being permitted which conserve energy. However out of all conceivable changes which satisfy this law there are many which do not occur in practice.

This is a remarkable observation. Although it is concerned with the first law of thermodynamics, it suggests that there is a special relation between the laws of a theory and the possible transitions from one state of a theory to another (the transitions from one state to another have to conserve energy).

There is one drawback to this idea. There are many scientific laws that don't concern the transition from a state at one time, to another at a different time, but concern what is going on in states at any one given time. A paradigm example: Ohm's law which concerns the connection between current and voltage in closed circuits.

**(2.4)** R. Giles offered an important account of states of a system that stressed how the states of any useful physical theory represent what he calls the preparation of systems which the theory studies, and he discusses when two states are identical.

> ...in a useful physical theory the *state* of a system represents its method of preparation: two systems are in the same state if they have been prepared in the same way, or more precisely *if our information about the method of preparation* is the same in each case.[12]

Clearly the characterization of a state as a representation of its method of preparation is a more definite, but somewhat idiosyncratic characterization of the state of a (thermodynamical) system than usual. Aside from that, Giles makes the interesting comparison of states and systems, in that there is a composition or union of systems (which he calls conceptual), For systems A and B, there is the primitive concept of their union, (A o B), and there is also a primitive "addition" of states, described this way

> ... *if a and b are two states then aob is the state whose method of preparation consists in the simultaneous and independent performance of the methods of preparation corresponding to the states a and b.* The associative and commutative laws of addition of states are evidently still applicable.[13] (Giles 1964, p. 22).

This observation by Giles adds a very important feature of states to their characterization: states can be added to form other states, and the operation of addition is

[11] *The Elements of Classical Thermodynamics*, Cambridge University Press, 1957, p. 29.

[12] *Mathematical Foundations of Thermodynamics*, 1964 Macmillan, p.17.

[13] Giles, 1964, p.22.

associative and commutative. The importance of Giles' observations lies in his making evident that the collection of states has a structure: not only can they be added, but there is a relation among states that seems to track a certain evolution, posssibly probabalistic, from one state to another. As he puts it,

> Let a and b be two states. We write b $\gtrsim$ a if there exists a state k and a time $\tau$ such that aok evolves (in isolation) in the time $\tau$ into the state bok. That is, ... such that the state whose method of preparation is "apply simultaneously and independently the methods of preparation corresponding to a and k and wait for a time $\tau$" is indistinguishable from the state whose method of preparation is "apply simultaneously and independently the methods of preparation corresponding to b and k," in the sense that any experiment applied to these states will yield the same result (or rather the same statistical distribution of results) in each case.[14]

Thus the states have a structure, and a certain relation among them corresponds to or represents a law of the relevant theory. We think that these apparently diverse accounts of states and laws reveal a general picture that we will try to explain below. We will leave to one side those accounts that tell us about states of systems in the cases of statistical mechanics and information theory.[15]

## 10.3 The Magnitude Vector Space of a Theory (MVS)

We now turn to characterization of a notion of the state of a theory which differs in detail from those that we have just surveyed. Nevertheless, we believe that this notion of state plays an important role in specifying critical information about the various systems studied in each theory. It also plays an important role in in explaining how theories and their scenarios set the stage for modal *physical* possibilities, and also helps to describe a feature of laws - how they determine a subset of the physical possibilities that the theory provides, which we will call *nomic* possibilities.

The touchstone for our account of laws rests on the following observation of laws and magnitudes that is based on H. Whitney's seminal account of the algebra of physical quantities[16] Here is what Krantz, Luce, Suppes, and Twersky say about that neglected seminal work:

> It is widely agreed that physical quantities combine additively within a single dimension (at least when that dimension is extensively measurable) and that different dimensions combine multiplicatively; that the multiplicative structure is very much like a finite-dimensional vector space over the rational numbers; that the existence of basic sets of

[14]Giles, 1964, p.24. p.112).

[15]Cf. *Foundations of Measurement, volume I, Additive and Polynomial Representations*, D. H. Krantz, R. Duncan Luce, Patrick Suppes, and Amos Tversky (1971, Academic Press), 2007, Dover Press. vol 1, p. 112.

[16]"The mathematics of physical quantities. Part I: Mathematical models for measurement. Part II: Quantity structures and dimensional analysis". *American Mathematical Monthly*, 1968, 75, 115–138, 226–256.

dimensions in terms of which we can express the remaining dimensions corresponds to the existence of finite bases in the vector space; and that *numerical physical laws are almost always fomulated in terms of a special class of functions defined on this space* (emphasis added).[17]

We agree with the assessment of Suppes *et al*, that most *numerical* laws are formulated in terms of a special class of functions on a vector space. In what follows, (Sect. 5 below) however we will propose a different vector space, the *physical magnitude vector space of a theory*, **(MVS),** rather than the one that Whitney proposed.

When we discuss laws and the magnitudes that are used in their description, we will generally take magnitudes to be functions that map objects to real numbers (or to mathematical structures like vectors, tensors, and fields of various kinds). In addition to those magnitudes, we think that certain *functionals* also have an important connection with laws. Functionals are functions of functions whose values are real numbers.[18] We will therefore also regard some functionals as physical magnitudes. Though they are not mappings of objects to the real numbers, but are instead mappings of abstract objects (functions) to the reals. Here are some familiar examples:

1. Let f be any magnitude that is a function from objects to the real numbers. Subject to certain conditions, the integral of that function f, over a closed interval of the real numbers from a to b, yields the area under that curve over that interval. No one disputes that under those conditions that area is also a magnitude.
2. If To(x) is the kinetic energy of a specific system of bodies, and Vo(x) is its potential energy at some time, then the integral with respect to time, from the time to, to the time t1 of the difference (To − Vo) (the *Lagrangean* of the system) is called the *Action*, and it is also regarded as a physical quantity. Since it is an important feature of our notion of a magnitude vector space that it is constructed from the magnitudes of a theory or theoretical scenario, it is important to note that these functionals are also included among the magnitudes.

As we have noted above, there is a wide variety of views on states of systems and theories. Nevertheless, we think that laws have an important relation to the states of theories. We shall now try to express those insights more sharply.

[17] *Foundations of Measurement, volume I, Additive and Polynomial Representations*, D. H. Krantz, R. Duncan Luce, Patrick Suppes, and Amos Tversky. (1971, Academic Press), 2007, Dover Press, p.459.

[18] The functionals on the elements of a vector space V are members of the dual vector space V^. There is an important connection between functionals on a magnitude space and laws which we shall explain in a later chapter.

## 10.4 A Shift: The Focus on Laws with Background Theories (or Theoretical Scenarios)

We begin with an important qualification. We shall now focus only on those laws that have some association with theories or theoretical scenarios, and we will neglect the orphan laws - those laws, which have no theoretical home. This shift may result in some loss, by narrowing the scope of our account, but we think it is worth pursuing. We think that there are two factors that favor this shift. The first is that it enables us to introduce a notion of possibilities that are *physical modals* and to distinguish among them a certain subset of *nomic modals* that are associated with laws.

The second factor in favor of this shift in focus is not entirely new. The requirement for a theoretical scenario or background theory has a distinguished ancient ancestry that goes back to a distinction first introduced by Aristotle - the distinction between knowing facts, and knowing reasoned facts. Aristotle thought that it was only a special kind of deduction that provided an understanding of laws. The special kind of deduction yields a reasoned fact. He says:

> Understanding the fact and the reason why, differ first in the same science, – and in that in two ways: in one fashion, if the deduction does not come through intermediates (for the primitive explanation is not assumed, but understanding of the reason why occurs in virtue of the primitive explanation); in another, if it is through immediates but not through the explanation but through the more familiar of the converting terms. For nothing prevents the non-explanatory one of the counterpredicated terms from sometimes being more familiar, so that the demonstration will occur through this.[19]

Laws then, are not just facts, they are reasoned facts. What makes them reasoned facts is that they are deduced from other generalizations. However, not just any deduction will serve. He also required that certain terms in the deduction are more familiar than others, and that requirement seems to be connected for him with the crucial element: the deduction is an explanation of the law.

One example that he gave of the difference between those deductions that yield just facts, and those that yield reasoned facts involves the deductions of the fact that the planets are not near, and deduction of the fact that the planets don't twinkle. The two deductions (I) and (II) below concerning astronomical objects, are modified slightly:

(I) (1) Objects are near if and only they do not twinkle.
 (2) Planets do not twinkle. Consequently,
 (3) Planets are not near.

(II) (1) Objects are near if and only if they do not twinkle.
 (2) Planets are not near. Consequently
 (3) Planets do not twinkle.[20]

[19]*Aristotle's Posterior Analytics*, Oxford University Press, 1975, Tr. J. Barnes, PA, A13. 78a23-78b15., and quoted in David-Hillel Ruben, *Explaining Explanation,* Routledge, London, p. 105.

[20]David-Hillel Ruben, *Explaining Explanation,* Routledge, 1992, p. 106.

The first argument according to Aristotle is a case of a deduction of fact, while the second is a deduction of a reasoned fact. The question is what makes the difference. There is a hint later in the text, where he mentions another example of the distinction:

> Another example is the inference that the moon is spherical from its manner of waxing. Thus: since that what so waxes is spherical, and since the moon so waxes, clearly the moon is spherical. Put in this form, the syllogism turns out to be proof of the fact, but if the middle and major be reversed it is proof of the reasoned fact; since the moon is not spherical because it waxes in a certain manner, but waxes in such a manner because it is spherical.[21]

Laws then, on Aristotle's account of them, are the deductive conclusions of a specific kind of syllogistic inference, where that deduction is explanatory. Thus on Aristotles' account of them, every law has a deductive explanation. So it's a very old tradition (or at least a very early precedent), that laws are different from contingent generalizations in that the laws require some explanatory deduction of them from suitable premises.

Our present proposal however is a weaker version of the Aristotelian idea. For us, sometimes the theoretical scenario of a law can provide an explanation of it. But not always. As we shall see (in Chap. 12), in those cases when the background is a schematic theory, there are significant scientific laws that are special applications that are subsumed under those theories. These laws count as successes of the schematic theories, but although they are *subsumed*[22]under those theories, they are not deductive consequences of them.

Given the magnitudes of the theory or schematic scenario that is associated with a law, we can then construct an associated vector space whose members are *states* of the theoretical background. We will propose that these states are the *physical possibilities* provided by the theoretical background of the theory.

## 10.5 The Magnitude Vector Spaces Associated with Theories

It is apparent from our brief survey, that no single notion of state is common to the wide variety of theories we have mentioned. Nevertheless we will define a kind of vector space that will be available for theories, and show its importance for understanding how laws are related to the states, – the modal possibilities, provided by these vector spaces,

Let us suppose then that each theory has a finite or possibly infinite number of magnitudes M, M1, ..., Mn, .... Since we want magntudes to include velocities, fields, tensors, etc., they have to be functions mapping to mathematical structures of various kinds, and the ranges of these functions can be different kinds of mathematical, and physical objects or entities such as fields, shapes, etc.

---

[21]Aristotle, Posterior Analytics, in Prior Analytics and Posterior Analytics, tr. by A.J. Jenkinson and G.R.G. MurePart, Digireads.com Publishing, 2006, §, slightly modified.

[22]We shall give a precise definition of the relation of subsumtion in Chap. 12.

It is usual, in referring to quantitative structures (as H. Whitney did), to think of magnitudes as having several basic operations on them. Here, for present purposes, we shall use only two: addition and multiplication.

(1) We have an addition operation (+), restricted to magnitudes of the same type:

$$rM + sM = (r + s)M, \text{and}$$

(2) There is a product operation (represented by concatenation) rμ, where r is any real number, and μ is any magnitude, and their product is again a magnitude of the same kind, such that for any magnitude M, and real numbers r and s, we assume that

$$r(s\,M) = (rs)M.$$

(3) There is also a kind of multiplication and division of the magnitudes themselves, though this has to be treated with special care since it may not be the case that every product of magnitudes is a magnitude, nor the division of any magnitude by another may not be a magnitude (sometimes yes: mass divided by volume is density – a magnitude. But, for example, length divided by time is not.

We shall not need the product and division of magnitudes in what follows, though they play an important role in dimensional analysis. Whatever else may be assumed about magnitudes in general, our assumptions (1) and (2) are always taken to be features of magnitudes, and not out of the ordinary.

We take very seriously, the insights mentioned above, of Whitney, Krantz, Luce, Suppes, and Tversky, that numerical physical laws are almost always formulated in terms of a special class of functions defined on a special vector space of magnitudes of which phase and configuration spaces are special cases. We do differ however from them in the kind of vector space that we have in mind.

We want to define a special type of vector space associated with each theory (or theoretical scenario), by showing how the basic magnitudes of a theory (or scenario) can be used to construct a vector space which we will call the *magnitude vector space* (**MVS**) of the theory. This space can be introduced by generalizing from what is essentially the phase space of any theory that can be represented by a Hamiltonian. Since those cases for which a Hamiltonian function can be used to cover a considerable amount of physical theory, it is a good place to begin.

Classically, for phase spaces in the Hamiltonian case, the Hamiltonian function H(p1, ..., pn, q1, ..., qn) for n particles consists of a set of ordered 2n-tuplets for any n-bodied system, where the n pis correspond to what is called the generalized momenta of each of the n particles, and the corresponding n qis are their generalized positions. These are usually regarded as the basic magnitudes of the theory. The dimension of these vector spaces, when finite, are obviously even in dimension (2n).

To generalize this particular way of looking at things, we start by indicating what the vector space is in the Hamiltonian case. The construction exploits the use of the magnitudes of the theory. We follow H. Whitney's usage and shall say that for each of the magnitudes pi (qi) we have the *rays* |Pi|, |Qi|, where

$$|\mathrm{Pi}| = \{\mathrm{rpi}|\text{for all real numbers r}\}, \text{and } |\mathrm{Qi}| = \{\mathrm{rqi}|\text{for all real r}\}.$$

We take the corresponding Cartesian product to be the phase space[23]:

$$|\mathrm{P1}\ \mathrm{x} \ldots \mathrm{x}\ |\mathrm{Pn}|\ \mathrm{x}\ |\mathrm{Q1}|\ \mathrm{x} \ldots \mathrm{x}\ |\mathrm{Qn}\ | \ .$$

We can now generalize this construction to cover cases of theories and their scenarios, whether they fall under the Hamiltonian case or not (for the present, we restrict our generalization to cases where there are only a finite number of magnitudes under consideration):

1. Consider any theory T (or theoretical scenario) whose basic magnitudes are M1, …, Mn (so, in this generalization we bypass a constraint that would limit us to symplectic manifolds which are always of dimension 2), and form the rays |M1|, …,|Mn|. Let the Cartesian product of these rays constitute the magnitude vector space of T. I.e.

$$V_{M,T} = |\mathrm{M1}|\ \mathrm{x} \ldots \mathrm{x}\ |\mathrm{Mn}|$$

that is, the set of all n-tuples (α1M1, …, αnMn), for all real numbers α1, …, αn. This set is a vector space, provided that we define two conditions such that for all ai and bj,

(4) (**Addition**) (α1M1, …, αnMn) + (β1M1, …, βnMn) = ((α1 + β1)M1, …, (αn + βn)Mn), and
(5) (**Multiplication**) γ(α1M1, …, αnMn) = (γα1M1, …, γαnMn) for all real numbers γ, and αi.

The effect of these operations on the vectors, is to preserve the type of magnitude in each component.

We now define a *state of the theory T* (or *theoretical scenario*) as any vector of the magnitude vector space of T – i.e. any vector <α1M1, …, αnMn>, where the αs are all real numbers.

The various systems that the theory can describe are "located" in this magnitude space by specifying the state or states it is in at various times. Thus the states of any system are the states provided by the theory. These states, we wish to say, represent the various *physical possibilities* for the systems under discussion. We believe that these possibilities are serious modal possibilities and not just a way of speaking

---

[23]The Cartesian product of two sets A and B is the set of all the ordered pairs <x, y> with x and y being members of A and B respectively.

modally with the vulgar. Showing that these possibilities are modal possibilities is the task of the next chapter.

However, the sense in which these states are the *possibilities* provided by the theory T has to be explained; it is not merely a loose way of speaking modally, but a serious kind of modal notion, which we shall call a *physical modal.*

We therefore turn, in the next chapter, to a more detailed explanation of this special kind of modal, in order to argue that it is the very states, the vectors of the magnitude vector space of a theory,[24] that are the theoretical possibilities, about which laws have much to tell us.

[24]That is, we think that the very states themselves that are modal possibilities, (analogoous to the idea that it is the elements of a probability sample space that indicate the possibilities). Each of these states is a modal possibility in a structure. There are examples of what we call *modal entities*, rather than modal operators. We also will show the connection of these possibilities to an associated modal operator.

# Chapter 11
# The Possibilities That Theories Provide (Physical Modals) and the Possibilities of Laws (Nomic Modals)

In the previous chapter we agued that every theory provides a collection of physical possibilities, – they are the (basic) elements of its associated magnitude vector space. The elements of that space are the *modal physical possibilities* provided by the theory. We will now explain why they are genuine modal possibilities, and not merely a loose way of speaking.

Before we proceed, it is important to observe an elementary distinction: there are at least three kinds of things one might think of as modal: modal ***predicates***, modal ***operators*** (these will be functions ***on** implication structures,* a key concept which we will presently define)) and modal ***entities*** (items that might be concrete or abstract that are ***in*** implication structures). We will be concerned in what follows only with modal operators and modal entities, and will try to make the difference clear as we proceed.[1]

## 11.1 Theoretical (Physical) Possibilities

So, there are then three tasks before us: (1) what kind of possibilities are these theoretical physical possibilities? Are they genuine modal possibilities or only a *façon de parler*?, (2) What role do laws play with respect to these possibilities? and, lastly, (3) what are the *nomic possibilities* – a class of modals that are a subset of the theoretical physical possibilities, and have a special connection with laws.

[1]In anticipation: we will not say anything about modal predicates, but we will study in detail a host of modal operators and with the help of them we will define the important case of modal entities such as the possible outcomes in a probability space, the magnitude states of theories and so forth.



A. Koslow, *Laws and Explanations: Theories and Modal Possibilities*, Synthese Library 410, https://doi.org/10.1007/978-3-030-18846-7_11

We will begin with (1). The example of the vectors in the magnitude space of a theory is an example of a kind of modal possibility that has elsewhere been called *natural possibilities*.[2]

To convey an idea of this special kind of modal possibility, we would like first to give a broad survey of examples of this kind of possibility, and then characterize this kind of possibility more formally using the usual notation of the box and the diamond. We will then argue that the (basic) elements of the magnitude space of a theory are akin to the more familiar case of the possible outcomes that are the members of a probability sample space.

Here then are some examples of the kind of modality that we have in mind:

(i) *Probability sample space possibilities*. A die is thrown and there are, as we usually describe them, six outcomes which are the possibilities. If there were two dice there would be 36 distinct possibilities. In this case the possibilities are just the elements of a probability sample space.[3]

(ii) *Truth-value possibilities*. Assuming bivalence, declarative sentences are described as being either true or false, and in the case of classical logic, we say that these are the only two possibilities. So, the truth-values are also possibilities.

(iii) *Possible world possibilities*. In the case of some versions of possible world semantics, it is assumed that a certain collection of worlds is such that it contains all the worlds, and that they are mutually incompatible. The usual terminology naturally describes such a case as a set of possible worlds, with each world being a possibility.

(iv) *Physical possibilities*. For most physical theories there is an associated notion of the states of that theory. The set of all of these is commonly described as setting forth the physical possibilities for those systems that are studied by the theory.

(v) *Configuration space possibilities*. Suppose that in a configuration space of some physical theory, two points A and B are distinguished. Usually it's said that there is an actual path or curve along which the system passes from state A to state B. All the other curves connecting A and B are described as possible routes or paths or orbits from A to B. Sometimes then, it is the paths which are

[2]A. Koslow, "Laws, explanations and the reduction of possibilities" in *Real Metaphysics Essays in Honour of D. H. Mellor*, Eds. Hallvard Lillehammer and Gonzalo Rodriguez-Peryra, Routledge, London and New York, 2003, pp. 169–183, where they were called Natural modals. in A. Koslow, *A Srtucturalist Theory of Logic*, Cambridge University Press, 1992, pp. 239–371, where they were called just modals, and A. Koslow, "The Implicational Nature of Logic", in *European Review of Philosophy, The Nature of Logic*, volume 4, ed. Achille C. Varzi, CSLI Publications, Stanford, 1999, pp. 111–155.

[3]Technically, we will use the Kolmogorov axiomatization of probability, so that the bearers of probability are sets. Though the outcomes in the case of a thrown die are 1, 2, ..., 6, their probabilities are the probabilities of their unit sets {1}, ..., {6} respectively. The reason is that probabilities are defined on Boolean fields, and they are sets of sets. So when we refer to the possibilities in the sample space, they are the unit sets of each of the outcomes.

the possibilities. This is an interesting example where even though the possibilities in some sense exclude each other, they also exist together, side by side (as it were).[4]

(vi) *Possible cases, in proofs by cases*. There are mathematical proofs that proceed by cases, in which case it is natural to say that there are two or more possible cases.

These examples of possibilities are very varied. They can be the outcomes of an experiment (or trials) as in the case of a probability sample space, or cases of a sentences being true (or false), or particular worlds used in a possible-world semantics. They can be as abstract as subcases in a proof, or paths in a state space on the one hand, or as concrete as paths in a wood for another.

## 11.2 Gentzen Implication Structures, and Their Modals

We think that despite their variety these modals that concern possibilities are modalsof a special kind – close to what are familiarly known as C.S. Lewis S5 modals. Furthermore, after defining these necessity and possibility operators, they will also be shown to belong to a large class of modal operators which we will term *Gentzen structural modals*- so-called called because they are defined with respect to any *Gentzen Implication structure* – which will be described below. First however, there is an immediate problem that has to be addressed: modal operators are usually defined and distinguished from each other with the aid of some implication relation.

It is a common view that implication relations can only hold between items that are sentences, schematic letters, truth-apt items, propositions, or objects of possible belief, among other things. If, sometimes the formulation of an implication relation uses sets, they are usually sets that have sentences, schematic letters, truth-apt items, propositions, or objects of possible beliefs as members. However, numbers, paths, and states of theories are none of the above, – and that would seem to be an impediment for considering modal operators on sample spaces. If that were so, then there would be no implication structures that related numbers, paths between places and states of theories. None of them could be related by an implication relation. Furthermore, if, as we believe, modal operators are always defined with respect to some implication relation already in place, then it is hard to see how one could talk of modal operators for numbers, paths, and states of theories.

These difficulties are imaginary. G. Gentzen[5] advocated a view of implication that permitted implication relations to hold even for items like events, and Tarski

[4]Poignantly described by Robert Frost's T*he Road not Taken*, "Two roads diverged in a wood, and I – I took the one less traveled by, And that has made all the difference."

[5]'Untersuchungen über das logische Schliessen" Mathematische Zeitschrift (1933), 39: 176–210, 405–31; reprinted (1969) as "Investigations into Logical Deduction," in, M. E. Szabo (ed.) The Collected Papers of Gerhard Gentzen, North-Holland Publishing Co., pp. 68–131.

occasionally spoke of the subset relation between arbitrary sets as an example of an implication relation. *Gentzen implication structures* can involve elements other than sentences, statements, propositions, or properties. They are very general. We hasten to explain how this is possible.

Let S be *any* non-empty set, and let $\Rightarrow$ be any relation on S, that satisfies the following six conditions which are the *Gentzen structural conditions* for implication:

(1) ***Refexivity***: $A \Rightarrow A$, for all A in S.
(2) ***Projection***: $A_1,A_2,\ldots,A_n \Rightarrow A_k$, for any $k = 1,\ldots,n$.
(3) ***Simplification*** (sometimes called Contraction): If $A_1,A_1,A_2,\ldots,A_n \Rightarrow B$, then $A_1,A_2,\ldots,A_n \Rightarrow B$, for all $A_i$ and B in S.
(4) ***Permutation***: If $A_1,A_2,\ldots,A_n \Rightarrow B$, then $A_{f(1)},A_{f(2)},\ldots,A_{f(n)} \Rightarrow B$, for any permutation f of $\{1,2,\ldots,n\}$.
(5) ***Dilution***: If $A_1,A_2,\ldots,A_n \Rightarrow B$, then $A_1,A_2,\ldots,A_n,C \Rightarrow B$, for all $A_i$,B, and C in S.
(6) ***Cut***: If $A_1,A_2,\ldots,A_n \Rightarrow B$, and B, $B_1,B_2,\ldots,B_m \Rightarrow C$, then $A_1,A_2,\ldots,A_n,B_1,B_2,\ldots,B_m \Rightarrow C$.

We follow Gentzen's idea of understanding these conditions as very general, so that *any* relation satisfying these conditions, on *any* non-empty set S, constitutes an implication relation on that set. Any relation that satisfies the six Gentzen structural condtions (1)–(6) will be called a ***Gentzen implication relation*** *on S.* And by a ***Gentzen implication structure*** I = <S, $\Rightarrow$) we mean any non-empty set S together with a (Gentzen) implication relation $\Rightarrow$ on it.

## 11.3 Necessity and Possibility Operators on the Vector Magnitude Space of Theories

We want to show that there are modal operators on vector magnitude spaces. To do so, we shall adopt a five-step strategy in explaining why we think that physical theories provide a set of possibilities that are genuinely modal possibilities.

### *11.3.1 Probability Spaces and Modality*

We consider first, the special case of a probability sample space that illustrates the way to construct an associated *Gentzen implication structure* for the probability example, and then proceed to make the case for any vector magnitude space of a theory. The strategy here is developed in five stages. We will first define modal operators for various structures, beginning with simple probability sample spaces as implication structures using the usual box and diamond notation, and in the fifth stage, show that all those defined operators on magnitude vector spaces are modal operators on those spaces.

**Stage 1** Using probability sample spaces, we will explicitly define a modal operator of necessity, and a modal operator of possibility (indicated by a box, and a diamond respectively) on that probability sample space. First, we define a Gentzen implication structure associated with a probability sample space.

Consider a simple probability sample space: the set, of the outcomes of a thrown die, $P = \{1, 2, \ldots, 6\}$. Let $P^*$ be the power set of P (the set of all, subsets of P). Take the implication relation on $P^*$ to be the subset relation. It is easily verified that the ordered pair, PROB = $<P^*, \subseteq>$, is a Gentzen implication structure on the probability space $P^*$.[6]

**Stage 2** We now wish to show that in the probability case, we can define necessity and possibility operators, $\Box$ and $\Diamond$, acting on $P^*$ (the power set of P). Then for all x in $P^*$, and the empty set $\varnothing$,:

$\Diamond(x) = \varnothing$, if x is *not* a singleton (singletons are the one membered sets of $P^*$), having only one member), and P otherwise.

So the diamond is a mapping from $P^*$ to $P^*$. The singletons of P are what we shall call the *basic possibilities of* $P^*$.[7] It easy to show that $\Diamond(x)$ satisfies the following conditions (we leave the proofs for the reader):

(i) $\Diamond(x) = P$ if and only if x is a singleton.
(ii) If x is a singleton then $x \Rightarrow \Diamond(x)$.

There is a kind of converse to (ii), namely (iii).

(iii) If $x \neq \varnothing$,and $x \Rightarrow \Diamond(x)$, then x is a singleton.
(iv) If $\Diamond(x) \Rightarrow x$, then x is not a singleton
(v) For any x and y in $P^*$, $\Diamond(x \cup y) \Rightarrow \Diamond(x) \cup \Diamond(y)$.
(vi) There are x and y in $P^*$, such that $\Diamond(x), \Diamond(y) \Rightarrow \Diamond(x \cap y)$ fails.

---

[6]We refer here to probability theory in the form given to it by A.N. Kolmogorov, which was heavily influenced by the way that Hilbert thought of axiomatized theories, *Foundations of the Theory of Probability*, second English edition Tr. edited by N. Morrison,added bibliography by A.T. Batucha-Reid, Chelsea Publishing Company, New York,1956, original German monograph published in *Ergebnisse Der Mathematik*, 1933. It is clear that Kolmogorov thought he was axiomatizing probability theory in a way that paralleled Hilbert's axiomatization of Geometry (1899), and he thought he was following Hilbert's suggestion, in 1900, that one of the outstanding open mathematical problems included the axiomatization (Hilbert-style) of probability and mechanics. We return to a more detailed discussion of Kolmogorov's theory of probability in Chap. 12.

[7]The use of the term "basic" is intended to call attention to the similar case where the set of possibilities is given by the sample space of a probability space, and the singletons of the elements of such spaces are usually called the elementary events of the space.

We will also define a necessity modal operator on $P^*$, as follows: for all x in $P^*$

$\square(x) = P$, if $x = P$, and

$\varnothing$, otherwise.

The necessity operator can now be shown to have the following properties:

(vii) "$\square$" is a necessitation operator: i.e. if x is implied by all the members of $P^*$ (i.e. it is a thesis), then $\square(x)$ is also implied by all the members of $P^*$ (it is a thesis).
(viii) "$\square$" is a T-modal operator– that is, for all x in $P^*$, $\square(x) \Rightarrow x$.
(ix) "$\square$" is a K4-modal operator – that is, for all x in $P^*$, $\square(x) \Rightarrow \square\square(x)$.
(x) "$\square$" is an S5-modal operator – that is, for all x in $P^*$, $\diamond(x) \Rightarrow \square\diamond(x)$, and the dual, $\diamond\square(x) \Rightarrow \square(x)$, holds for all x in $P^*$.
(xi) $\square(x) \Rightarrow \diamond(x)$ does not hold for all x in $P^*$ (P is a counterexample).

However, the restriction of (xii) to basic events (i.e.singletons) does hold. i.e.

(xii) $\square(x) \Rightarrow \diamond(x)$ for all basic events.

Moreover, there are these partial results:

(xiii) $\square(x) \Rightarrow \neg\diamond(\neg x)$ holds for all x, (the converse fails for the empty set), and
(xiv) $\diamond(x) \Rightarrow \neg\square(\neg x)$ holds for all x in $P^*$ (the converse fails when x is P).

With the result (xii), this modal operator parts company with the many modal systems for which the box always implies the diamond. Nevertheless, for the items that we have termed the basic possibilities (the singletons of $P^*$), the box implies the diamond for every basic event, but fails for non-singletons. This result becomes evident once it is realized that the box is a T-modal, but the diamond is not. (i.e. (iv)).

Moreover we have with (xiii) and (xiv), only a partial overlap because only half the usual classical equivalences between the box and the diamond are available here. The box always implies not-diamond- not; but not conversely, and the diamond always implies not-box-not; but not conversely.

For some, the failure of the box to always imply the diamond (even though it does so for the basic events, (the basic possibilities), may seem like a failure of of modality for the operators described. We do not think so. With the exception of (xii) the modals described here comes close to being C. I. Lewis' S5.

It is not at all damaging that the box fails to imply the diamond for the basic possibilities. There are many kinds of familiar modals that share this feature. Perhaps the best known of those modals which break the uniform connection between necessity and possibility, is the Gödel-Löb model, $\square_{GL}$ where, roughly speaking, "$\square_{GL}(A)$" says that A is provable in Peano Arithmetic. It is well-known that for this modal, $\square_{GL}(A) \Rightarrow \diamond_{GL}(A)$ doesn't hold for every A. The reason is that for any thesis $\Omega$ of Peano Arithmetic, $\square_{GL}(\Omega)$ is also a thesis since (i.e. is implied by every element of the implication structure) the Gödel-Löb modal is a necessitation modal. In that case however $\diamond_{GL}(\Omega)$ is also a thesis. That however is impossible since no statement of the form $\diamond_{GL}(A)$ is a thesis.

In contrast, the diamond in the case of this probability sample space is a thesis for every one of the basic events (where the thesis of any implication structure is an element of the structure that is implied by every element in the structure), since the diamond of any singleton is S, which is a thesis, whereas in the Gödel-Löb case, the diamond of A is never a thesis.[8]

So far, we have defined modal operators of possibility and necessity on the set of this particular implication structure. They are operators on the set of elements of the implication structure. We also think of the singletons of the structure as its possibilities. These possibilities are elements of the set, and as such, they are possibilities **in** the structure. The important distinction between the modal possibility and modal necessity operators ***on*** an implication structure, and those modal entities (which are not operators), ***in*** an implication structure will be explained in more detail in stage 5.

**Stage 3 The necessity and possibility operators, $\Box_T$ and $\Diamond_T$, on vector magnitude spaces of theories.** At this stage we can now define modal operators on the magnitude spaces of physical theories in a way that is exactly similar to the way we defined necessity and possibility operators for the probability case. The importance for us will be this: for any physical theory, we will now have a way of understanding how it is that each physical theory can provide necessity and possibility operators associated with it.

We begin by associating with each theory T, a Gentzen implication structure:

Let S be the set of vectors of the magnitude space of a theory T, and let $S^*$ be the power set of S – i.e. the set of all subsets of S. We can then associate with the physical theory T, a Gentzen implication structure given by the ordered couple consisting of the set $S^*$,and the implication relation on $S^*$ given by the set-theoretical relation of the subset relation on sets. I. e.

$$I_T = \langle S^*, \subseteq \rangle$$

Then we define the necessity and possibility operators on the structure T as follows:

$\Diamond_T(x) = \varnothing$, if x is *not* a singleton and
S, otherwise.

$\Box_T(x) = S$, if $x = S$, and
$\varnothing$ otherwise.

These definitions for the modal possibility and the modal necessity operators are similar to what were provided for the probabilistic case – only the implication

[8]Cf. G. Boolos, *The Unprovability of Consistency*, Cambridge university Press, 1979, Theorem 11, p. 31.

structures are different. Note that in this case as in the probabilistic case, the singletons ***in*** the magnitude space of a theory are its modal possibilities, and are different from the modal possibility operators ***on*** the implication structure. As with the probability case, the important distinction between a modal possibility operator ***on*** an implication structure on the one hand, and the modal possibilities ***in*** the implication structure, on the other, will be deferred to Stage 5.

**Stage 4** Thus far we have provided explicit definitions for the modal operators on various structures. As modal operators, they are close to being S5, and the proofs for the features that they have, and those that they do not ((i) – (xv)) are straightforward.

Here is a convenient place to define a concept intimately related to implication structures: *Gentzen modal operators.* These modal operators will play a role in the development of our theory of the modal operators, and modal entities of physical theories and laws. In particular, it will enable us, with the aid of implication structures, to study the difference between the modal operators, and the modal entities of those structures. The Gentzen structural modal operators are defined as follows:

Let S be a non-empty set with a Gentzen implication relation ⇒, on it. We shall say that φ is a *Gentzen modal operator* on S if and only if it is a function on S, with values in S, such that the following two conditions hold:

(I) If for and sentences A1, A2, . . ., An, and B in S, such that A1, A2, . . ., An ⇒B, then φ(A1), φ(A2), . . ., φ(An) ⇒ φ(B), and
(II) For some A and B in S, φ(A ∨ B) fails to imply [φ(A) v φ(B)].[9]

[9]We regard the familiar bread and butter modal operators, and the Gentzen modals in particular, as only modal with respect to some implication relation on a non-empty set. This kind of modal operator was studied extensively in A. Koslow and there is a lot of evidence that all the familiar modals are also Gentzen modals, and all the systematization of modal theory by Kripke can be recaptured by using Gentzen modals. These efforts can be found in A. Koslow, *A Structuralist Theory of Logic*, Cambridge University Press, Cambridge 1992, Part IV, pp. 239–371, and "The implicational Nature of Logic: A Structuralist Account", in *European Review of Philosophy, The Nature of Logic,* volume 4, edited by A. Varzi, 1999, CSLI Publications, Stanford, pp. 111–155, where it is simply called a modal operator. The second clause of the definition is used for the special case when there is a disjunction operator always available in the implication structure. That is not always the case. The general condition for (II) is given by using the dual of the implication relation i.e. ⇒^ which always exists in implication structures in which there is an implication relation ⇒. It is defined this way: A1, A2, ..., An ⇒^B if and only if for any C in S, if all the Ai s imply (⇒) C, then B implies (⇒) C. In structures in which there is always a disjunction operator available, this is equivalent to the result that the Ai s together dually imply B if and only if B implies the disjunction of all the Ai s. In the single premise case, this comes to A ⇒^ B if and only if B ⇒ A. The remarkable discovery of the dual of an implication relation is due to R. Wojcicki, "Dual Counterparts of Consequence Operations", presented of *Seminar of the Section of Logic, Polish Academy of Sciences*, December 1972, pp. 54–57. Condition II in full generality requires that the operator does not distribute over the converse implication relation.

In the end we also want to show that these physical modal operators that we have explicitly defined, are *Gentzen structural modals*. Consequently all the physical modals we have introduced belong to a type to which most of our bread and butter modals belong. So, at the next stage, we offer some simple proofs to show that the new modal operators associated with theories and laws also belong to this wider group – the Gentzen structural modals.

To establish this connection we have to verify the two conditions for Gentzen structural modals on implication structures that are associated with a physical theory T:

(I) **Distribution over implication**: For any sets A1, A2, ..., An, and B in S*, we have

If A1, A2, ..., An $\Rightarrow$ B, then $\Box_T$(A1), $\Box_T$(A2), ..., $\Box_T$(An) $\Rightarrow$ $\Box_T$(B).

and

(II) **Non-distribution over the dual implication** ($\Rightarrow^\wedge$). There are sets Ao and Bo in $S^*$ such that $\Box$T (Ao $\cup$ Bo) does *not* imply $\Box_T$(Ao) $\cup$ $\Box_T$ (Bo).

The proof of (I) is as follows: Suppose that A1, A2, ..., An $\Rightarrow$ B. If all the Ai are S, then so too is B. Therefore all the $\Box_T$(Ai), and $\Box_T$(B) are S and so $\Box_T$(A1), $\Box_T$(A2), ..., $\Box_T$(An) $\Rightarrow$ $\Box_T$(B) holds. On the other hand, if for some k, Ak is different from S, then $\Box_T$(Ak) is the empty set, and so $\Box_T$(A1), $\Box_T$(A2), ..., $\Box_T$(An) $\Rightarrow$ $\Box_T$(B) holds.

The proof of (II): Let Ao and Bo be two sets of S* each different from the set S, and such that their union is equal to S. This can always be done. Then $\Box_T$(Ao $\cup$ Bo) is the set S. But $\Box_T$(Ao) $\cup$ $\Box_T$ (Bo) is the empty set because it is the union of empty sets. Consequently, the conclusion follows since S is not a subset of the empty set.

**Stage 5 Differentiating between "It is possible that A "($\Diamond$(A)", and "A is a possibility": Modal operators and Modal Entities.** We have described members of the spaces $S^*$, as (basic) possibilities if and only if they are the singletons of the appropriate implication structure. We noted in our discussion of the modals of the probabilistic implication structure that the singletons *in* that structure were the basic outcomes of the probability sample space. In a parallel way, we think that the singletons *in* the magnitude vector spaces of theories are the (basic) possibilities *in* the implication structures associated with physical theories They are not to be confused with possibility operators or with the values of possibility operators. They are modal entities – that is they are entities that are modal and are members ***in*** the set that is part of the implication structure.

By its definition, $\Diamond$(x), for any value of "x" can never itself be a possibility ***in*** the structure. The reason is that the value of the diamond operator for any value of "x", is either the set S or the empty set, and neither of these is a singleton. The case for the box operator is a little different. Sometimes (for some value of "x"), $\Box_T$(x) is not a necessity (i.e. it is sometimes the empty set for certain values of "x").

In short, the box and diamond *operators* are functions, and their values respectively are not necessities or possibilities in the implication structures. We have to be careful therefore about the way to express that something is a possibility (or a necessity) and to distinguish those facts from those which express the values of the operators.. They are very different from each other.

In order to express the fact that any x in the space S∗ (i.e. in the implication structure) is (or isn't) a (basic) possibility, and that x is (or isn't) a (basic) necessity in the structure, we will introduce two predicates: "Nec(x)", and "Pos(x)", that say, respectively, For some A and B in S, ,that x is a (basic) necessity, and that x is a (basic) possibility, respectively. They are defined as follows:

(1) x is a (basic) necessity ***in*** the structure, Nec(x), if and only if $\Box_T(x) = S$.[10]

"Nec(x)" has the nice property that if A implies B, then Nec(A) implies Nec(B). (Remember that for the set $S^*$, the implication relation is set inclusion). So if anything is a necessity, then anything it implies is also a necessity.

(2) x is a (basic) possibility, Pos(x), ***in*** the structure if and only if $\Diamond_T(x) = S$.

"Pos(x)" has the nice property that for any non-empty set A implies B, then Pos (B) implies Pos(A). So, for anything that is a basic possibility in the structure, then any non-empty set which implies it is also a basic possibility. Recall that

(3) x is a thesis of any structure if and only if every element of the structure implies x.

It follows from (1), (2), and (3) for the structure IT, that that

(4) There is only one thesis in the structure, namely S.
(5) There is only one basic necessity in the structure, namely S.
(6) There are many basic possibilities in the structure., and, perhaps most interestingly, unlike what we encounter in most modal systems,
(7) Nothing **in** the structure can be both a necessity and a possibility.

For if there is some element in the structure, say a, such that both Nec(a) and Pos(a), then $\Box_T(a) = S$ and $\Diamond_T(a) = S$. Consequently, a is both identical to the set S, and is also a singleton, and that is impossible (assuming of course that S has at least two elements).

[10]This has the nice consequence that the only basic necessity in this structure is the set S. In this structure S is the union of all the singletons of S. Consequently, in this structure the union of sets is their disjunction, so that the only necessity in this structure is the disjunction of all the basic possibilities. That conforms to the usual mathematical way of talking about the necessity of S. in such cases. I.e. in the probability case for example, what is the basic necessity in the structure is the disjunction of all the basic outcomes.

## 11.4 Magnitude Vector Spaces, and Their Theoretical (Physical) Possibilities

We now return to the results of Chap. 9, §3, on those Magnitude Vector Spaces associated with theories (or theoretical scenarios), in order to focus on the connection of laws with the states of the magnitude spaces of theories.

For the magnitude vector space of a theory T, we have a space whose members consist of possibilities that depend critically on the physical magnitudes of T. For that reason we think of the states of the associated vector spaces not just as the theoretical possibilities provided by physical theories, but as the *physical possibilities* provided by them. These possibilities can be different for different theories – but needn't be.

Thus far we have considered those features of laws that involve their relation to other *sentences*. Those sentences may occur in explanations of their instances, or may appeal to more inclusive theories that explained them, from which they deductively follow. Those features appealed to the truth of other sentences, or classes of such sentences. We will now consider some features of laws that depart from that tradition by examining the relation of laws to items that may not be sentential, – like nomic possibilities and linguistic items that may not even be true, such as schematic theories which are not only not true, but, as we shall see (in Chap. 12), they do not deductively imply laws that are associated with them.

We have already indicated how a theory can provide collections of physical possibilities that are modal possibilities. We wish now to explain how laws single out special subsets of those physical possibilities which we will call their *nomic* possibilities.

## 11.5 Laws and Law-Designated Possibilities: The Nomic Possibilities

Each law, as we shall see, tells us which physical possibilities are nomic, which physical possibilities are non-nomic, and, sometimes tells us how some nomic possibilities depend upon others – a kind of closure condition.

It is helpful to distinguish the nomic possibilities from among the physical ones with the aid of two classic simple examples: Ohm's law, and Galileo's Law of Falling Bodies. Ohm's Law is a very powerful tool in solving complex problems in circuitry and its scientific merit and usefulness is not in dispute. Its limited scope, however, outside of which it is unreliable, may, for some, diminish its use as a scientific law. Despite those drawbacks, it, along with Galileo's law of falling bodies are useful ways of introducing the notion of nomic possibilities. These modals are a subset of the physical modals associated with theories and theoretical scenarios, and they play an interesting role with respect to laws.

### 11.5.1 Ohm's Law

Georg Ohm introduced the law that $V = R \times I$ in 1827, and according to J.C. Maxwell,[11] Ohm simultaneously provided a theoretical scenario consisting of a group of new concepts that he used to express that law -ie "electromotive force" ("V" for volts nowadays, though Ohm used "E"), "strength of current (I), and "electrical resistance" (R) replaced the vague, non-quantitative concepts used prior to Ohm's paper. Thus I will regard it as a case where the theoretical scenario and the law are introduced in one fell swoop.

We can now illustrate what laws say about the states of their associated magnitude vector spaces. In the Ohm example the magnitude vector space is the one that is constructed using the two magnitudes V, and I. Ohm's law is usually represented by the eq. $V = I \times R$ (call it "H"). Consequently, we know that there is a function f, of the two magnitudes V and I, such that $f(V, I) = 0$ on the associated vector space. Now define a subset D(H) of the vector space as an H *-designated subspace* of the vector space if and only if it consists of exactly those pairs $< M1, M\ 2>$, such that

$$M_1 = M_2 \times R.$$

Given a particular law, say H, we shall simply say that D(H) is a *designated subset* (indexed to the law H) of the set of physical states or physical possibilities. The members of this designated set we now think of as *nomic possibilities* of H, the law under consideration. The nomic modals are a special subset of the physical possibilities, so they are, like the physical possibilities of a theory, modal entities.

For Ohm's law the designated set D(H) is given by:

(1) $< M_1, M_2 > \in D(H)$ if and only if $M_1 = M_2 \times R$,

where $< M1, M2 >$ is a state in the associated magnitude space.

It follows, that for the particular magnitudes V and I, and the constant R, that

(2) $< V, I > \in D(H)$ if and only if $V = I \times R$ (i. e. H).

From (2), assuming Ohm's law, H, it follows that *some* states (possibilities) are designated states of the theoretical scenario, i.e. $< V, I > \in D(H)$. Moreover, according to (2), the law implies that if $< P, Q >$ is a theoretical state such that $P \neq Q \times R$, then that possible state is *not* a designated state.

With this terminology in place, the law says that some *physical* possibilities are nomic possibilities, and some physical possibilities are not nomic possibilities.

---

[11] Cf. the incisive discussion by J.C. Maxwell "On Ohm's Law". Reprinted from the *British Association Report*, 1876, in *The Scientific Papers of James Clerk Maxwell*, ed. W.D. Niven, MA., FRS. Two volumes bound as one volume. Volume 2, pp. 533–537.

### 11.5.2 Galileo's Law of Falling Bodies

For the Galilean law, the Galilean designated subset D(G) of the magnitude space is given by:

(3) $< x, t > \in D(G)$ if and only if $x = (1/2)gt^2$.

A conclusion similar to the Ohm case follows: Condition (3) implies that some of the physical possibilities of the magnitude vector space are nomic (those in D(G), and it also implies that some physical modals of the vector space are not nomic (those that are not in D(G).

These two examples can be seen as special cases of the way in which nomic possibilities are associated with laws:

Let H(p,q) be a law that involves just the two magnitudes p, and q. Then let $V_H$ be the associated magnitude vector space. Next, define the H-designated subset of the possibilities of $V_H$ to be the set D(H) defined this way:

$$(D) < p, q > \in D(H) \text{if and only if } H(p, q).$$

That is, a possibility in the vector magnitude space $V_H$ is H-designated if and only if H(p,q). In this way those possibilities which are H-designated as *nomic* possibilities are singled out in the associated magnitude vector space

So far, we have associated nomic possibiliies to any law, *if* that law can be expressed as a relation between physical magnitudes. This association can be generalized to include any law that is a relation of magnitudes. How general would the generalization be?

### 11.5.3 A Mach-Style Generalization of (D)

The assumption that *all* laws are functions of one or more physical magnitudes would enable us to provide nomic possibilities for any law. This assumption seems close to a proposal about laws that was suggested by the physicist and philosopher Ernst Mach at the turn of the twentieth century. According to Ernst Mach the coverage would be universal.[12] We can reformulate his insight, the *Mach Condition* (**MC**) as follows:

If a law L involves the physical magnitudes M1, M2, ..., Mn, then there is a function FL, such that

(**MC**) $L(M_1, M_2, \ldots, Mn)$ if and only if $F_L(M_1, M_2, \ldots, Mn) = 0$.

[12]Mach claimed that "The laws of nature are equations between the measurable elements $\alpha\ \beta\ \gamma\ \delta \ldots$, $\omega$ of phenomena.", Ernst Mach, *Science of Mechanics*, Chicago, The Open Court Publishing Company, 1907, Tr by T. J. McCormack. Of course special care has to be taken in our understanding of that dictum, since what he intended by "elements of Phenomena, as well as his notion of a mathematical function may be non-standard, and not the same as any contemporary notion.

There are qualifications that have to be in place, for the notion of a function might be taken in such a wide sense that the Machian Condition could then force an over-generation of the number of relations that count as laws. Nevertheless, (**MC**) enables us to draw a consequence that extends (D) to laws generally. I.e.

If L(M1, M2, ..., Mn) is a law, then, assuming (MC), $< M1, M2, \ldots, Mn > \in D(H)$ if and only if $FL(M1, M2, \ldots, Mn) = 0$.

This tells us that if some sequence of magnitudes, the vector in magnitude space <M1, M2, ..., Mn) are such that the law L is true, then the corresponding possibility (a specific vector in the associated magnitudes space), is a designated L-possibility, i.e. a nomic possibility, and it is not nomic if it fails to be L-designated.

We should point out that a nice feature of the Mach Condition is that it is formulated in a way that does not require that laws have a certain logical form; however the law is formulated, (**MC**) requires only that there be a certain function of the magnitudes used in the formulation of the law.

### *11.5.4 A Logical Relation (Gentzen Implication Relation) Between Nomic Possibilities*

We have argued that for any law – say, H, the nomic possibilities are those physical possibilities that are H- designated -that is, they are the members of D(H). As such they are not the sort of things that are true or false, and it would be unusual to think that that there could be anything like an implication relation that holds or fails to hold among the possibilities -which after all are the members of magnitiude vector spaces. Nevertheless, we think that for any law, say H, there is a *Gentzen structural limplication relation,*[13] $\Rightarrow^{H}$, that can be defined between those states that are physical possibilities:

Let H be some law, and for any nomic possibilities s1 and s2, define a H-Gentzen implication relation $\Rightarrow^{H}$ as follows;

(**HI**) s1 $\Rightarrow^{H}$ s2 if and only if [either s1 is not H-designated or s2 is H-designated].[14]

It is easily checked that this relation satisfies the six conditions for being a Gentzen implication relation. It is also of some interest that this Gentzen implication relation guarantees that any H- Gentzen implication of a nomic possibility is also a nomic possibility. That is,

For any law H, and *physical possibilities* S1 and S2, if S1 $\Rightarrow^{H}$ S2, and S1 is a nomic possibility, then so too is S2.

[13]Cf. section (2) above.

[14]The case for multiple antecedents is defined in the usual way that requires that either one of the antecedents is not H-designated or the consequent is H-designated.

The reason is fairly direct. Suppose that S1 is <p, q>, S2 is <r,s>, and S1 $\Rightarrow^{H}$ S2. Suppose further that S1 is a nomic possibility. Then <p, q > is in D(H). Since S1 H-implies S2, we have that either <p,q > is not in D(H), or < r,s > is in D(H). Since <p, q > is in D(H), we have <r,s > is in D(H). Therefore S2 is a nomic possibility.

This is a nice feature of laws. For any law, H, there is a Gentzen implication relation $\Rightarrow^{H}$, such that being a nomic possibility is inherited from any nomic possibility that implies it.

Not much is known so far about the H-Gentzen implication relations, $\Rightarrow^{H}$ on the magntude spaces that are indexed to laws. Here is one interesting feature:

Consider Galileo's law of falling bodies: G(x, t): $x = 1/2\ gt^2$. It implies that if, at say time to, the distance fallen is $xo = 1/2\ gto^2$, then, for example, at a later time, say 2to, the distance traversed will be 4 times xo. More generally, it follows from G(x, t) that G(xo, to) $\rightarrow$ G($n^2$xo, (n to). Then we have

If <x, to> $\in$ D(G), then < $n^2$ xo, nto > $\in$ D(G), then
<x, to> $\Rightarrow^{G}$ < $n^2$ xo, nto > .

In other words, generally, if a law H(p,q) implies a transition from one vector V in the associated measurement vector space) to another V′ (i.e. from one nomic possibility to another) then they are related in that order by the Gentzen H implication relation $\Rightarrow^{H}$ .

There are several immediate obvious consequences of the condition (D). The first is that any two logically equivalent laws H and H∗ have the same designated sets. That is, if H $\Leftrightarrow$ $H^*$, then D(H) = D($H^*$). I.e.. equivalent laws have the same nomic possibilities.

The second is that there is a difference between a law H being vacuously true, and the condition that the set of nomic possibilities, D(H), is empty. The reason is simply that if H is vacuously true, then it is true. But if D(H) is empty, then H is false (by (D)).

The aim of this chapter was to indicate how to each theory there is associated a collection of possibilities. This was done by showing how each theory provides a set of possibilities – its magntude vector space. The members of that set are states of the theory, and it was argued that its basic elements are genuine modal possibilities. We then distinguished between modal operators and modal entities. The states are modal entities: i.e. they are members of the set in the associated implication structure

We then showed how each law that is associated with a theory helps to pick out the *nomic possibilities* of the law from among the physical possibilities of that theory. We also defined (using the modal operators), two predicates "Nec(x)" for "x is a (basic) necessity", and "Pos(x)" for "x is a (basic) possibility", which sort out those entities in the magnitude space that are the possibilities and necessities in that space.
These definitions have interesting consequences – eg. there is no x in the structure such that Nec(x) and Pos(x). I.e. no entity of the vector space of a theory can be both a necessity and a possibility.

In the following final Chap. 12, we will focus on an important, but neglected situation: the consideration of laws that have theoretical backgrounds or scenarios, where those background theories are *schematic*. These laws (such as conservation of energy, conservation of linear momentum, angular momentum, and even Newton's first Law of Inertia) are incorporated, gathered under, or as we prefer to say (as we did in Chap. 8), *subsumed* under schematic theories. We discuss in detail some examples of physical theories that are among the most powerful ones known: Newton's Theory of Classical mechanics, Lagrange's theory of mechanics, Hamiltonian theory, and Kolmogorov's theory of probability and propose that all of them are schematic theories. These schematic theories are accorded great scientific merit because of their successes- their subsumed laws – even though none of the subsumed laws are deductive consequences of them.

However, all is not lost. We noted at the end of Chap. 9, that although schematic theories don't have *deductive* explanations of their own, they can have statements subsumed under them which explain those laws that are subsumed under them. And as we noted in that chapter, factivity also holds in the sense that if a schematic theory subsumtively explains any A, then A is true. I.e. subsumptive explanations have the kind of factivity that "O knows that P" enjoys.

We also want to suggest that subsumtive explanations have another feature which has sometimes proposed for theoretical explanations, but which has not to my knowledge ever been adequately described: It is this: *theories should have greater generality than what they explain.* To my knowledge, there is no acceptable account yet of "more general than" that meets that requirement.[15]

Nevertheless, we think that there is a kind of generality that schematic theories have. They have "instances", – those special cases or applications that are subsumed under them. However, those instances are not deductive consequences of the schematic theory, and that marks a major difference from the greater generality that universal generalizations have over their deductive instances. However schematic theories have a relation one should call *greater subsumtive generality* with respect to the applications that are subsumed under them. Of course this is not the kind of greater generality that universal statements have with respective to their instances, and it is not the kind of greater scope generality whereby universal statements like " All animals are mortal" are more general than "All humans are mortal". It is a kind of greater generality whereby a schematic theory is more general than any of the applications that are subsumed under it

One of the things we will also explore in the following chapter, is a possible way in which this notion of subsumtive generality that schematic theories have, might be extended to those cases where non-schematic theories explain.

---

[15]The two attempts discussed in Chap. 8, fall short. One, that could be tacitly assumed by R. Braithwaite requires that the explanation of a law be carried out in a deductiive system that positively extends the deductive system that contains that law. It isn't explained why this would confer more generality on the explaining theory. The other, provided in explicit detail by E. Nagel generalizes the idea that one generalization like "All Americans are mortal" is less general than the generalization that all human beings are mortal. The proposal has counterxamples, and also allows cases where logically equivalent statements can differ in generality. No immmediate correction is evident.

# Chapter 12
# Schematic Theories, Subsumtion of Laws, and Non-accidental Generalizations

## 12.1 The Incorporation of Laws into Schematic Theories

There is a raft of issues that have to be taken into account when the background theory for a law is schematic.

Recall that we first introduced schematic theories in our reconstruction of Ernest Nagel's account of how theories explain facts and laws (Chap. 8). That particular use of schematic theories however, is not the end of the story. We also introduced them in our discussion of David Hilbert's proposal, (Chap. 9), that the proper axiomatization of any theory, physical or mathematical, is a *Fachwerk* of concepts. We argued, on the basis of the meaning of "Fachwerk", that his view was that *all* properly axiomatized theories are schematic.

We noted that in addition, Hilbert claimed that the concepts of those axiomstized theories are those items which are the *infilling* of the Fachwerk – the replacements for the various schematic letters in the theory, subject of course to the constraints required by the Fachwerk. This view of Hilbert's explains, in part, the great emphasis that he placed upon a proof of consistency – even for physical theories: One has to show that there are some replacements of the schematic letters of the theory such that the constraints of the Fachwerk are met – i.e. consistency was important – even for physical theories

I am not suggesting that Hilbert was right in advocating this grand thesis. Some scientific theories may not be schematic and nonetheless be part of serious scientific inquiry, and may even have serious scientific explanations to their credit. Our view is the weaker one that there are a significant number of schematic theories that are theoretical scenarios for important laws. This alone requires a rethinking of some well- entrenched views.

These schematic theories give some insight into those laws, revealing deep truths about space, time, and energy (for example). Yet those laws are not explained by those theories; not if all the statements in explanations have to be true.



A. Koslow, *Laws and Explanations: Theories and Modal Possibilities*, Synthese Library 410, https://doi.org/10.1007/978-3-030-18846-7_12

As we shall see, those laws are particular applications or instances of the schematic theories. They are not deductive consequences of them, but are subsumed under them. Moreover, schematic theories can't be part of any explanation of the associated laws that are subsumed under them, for the reason that schematic theories have no truth-value, and so the factivity condition on explanations fails. Nevertheless, the concept of explanation∗ that we introduced in Chap. 9 (*explanation by subsumption,* i.e. *explains∗*) of course does hold for schematic theories, and it has half the factive property: If any schematic theory T explains∗ E then E is true. Nevertheless, if this is right, then the claim that explanations of laws is always deductive, is not correct.

Being schematic is not an unusual claim for some mathematical theories, but it might seem to be a controversial claim for physical theories. We shall now try to make the case that some familiar, very powerful physical theories are schematic. For example: (i) The Newtonian theory of Mechanics, (ii) Lagrangian theory, (iii) Hamiltonian theory, and (iv) Probability theory (the classic Kolmogorov version). We turn next to explaining why we believe that these familiar theories are schematic.

### *12.1.1 The Newtonian Theory of Mechanics*

By the *Newtonian Theory of Mechanics*, we mean Newton's three laws of motion. The first is his Law of Inertia. If there are no forces acting on a body, then it doesn't accelerate.[1]The Second Law is, roughly stated, that $F = m \times a$ (the force on any body equals the product of its mass and its acceleration (where force and acceleration are vector quantities and mass is a real-valued scalar quantity). The third law is usually expressed as requiring that every action has an equal and opposite reaction. It however will not play a role in what we wish to say about his theory of mechanics. Newton's Law of Gravitation however, is not part of his theory of mechanics, but part of his theory of gravitation (Book III of his Principia), and will be discussed below.

In representing the Second Law as an equation, "$F = m \times a$", it is clear that we have departed from Newton's original formulation of it, in at least two ways. The first is that we use equality in stating it, and he used ratios. The second, more important difference involves the mathematical representation of forces. Whereas Newton represented forces by appropriately directed line segments placed in

[1]We have already noted above a reason why the first law does not say what happens when the total force is zero. Rather, it says what happens when there are **no** forces on a body. We argued previously that the total force interpretation is not a correct expression of the law. Here is J.C. Maxwell's similar way of representing the first law: "The first law tells us under what conditions there is no external force." (p. 27) *Matter and Motion, with notes and appendices by Sir Joseph Larmor.* Original edition 1877. Dover Publications, Inc.,

geometrical diagrams, we follow the later mode of representation of forces by mathematical functions.[2]

We do however follow Newton closely in several ways: We follow him in his in his description of the First Law as concerned with what happens when there are *no* external forces acting on a body; not that the total external force is zero. More important, we agree that the Second Law does not imply the First Law. Newton was clear about this difference, and we follow him in that matter.[3] Newton was a superb mathematician. If the Law of Inertia was an easy consequence of the Second Law, as some present accounts have it, that fact would not have escaped him.

Most important of all, we adhere to Newton's presentation of his Second Law. We believe that it concerns the connection between *any* force and its associated acceleration. The Second Law is not restricted to the consideration of the special case of the *total external force* on a body and the body's associated acceleration.

Here's how Maxwell in particular stated it:

> When any number of forces act on a body, the acceleration due to each force is the same in direction and magnitude as if the others had not been in action. (37)[4]

Maxwell's use of the subjunctive was meant to express the fact that the resulting accelerations associated with each force are independent of each other.[5] The idea is that for each external (impressed) force on a body, there is an associated acceleration, and each one of the forces on a body is the product of its mass and acceleration. [6]

---

[2]However, some authors such as J.C. Maxwell, and R. Feynman, use both kinds of representation.

[3]There is an interesting argument why the Second Law does not follow from the first that is due to I.B. Cohen and Anne Whitman, *Isaac Newton, The Principia A New Translation, Preceded by a Guide to Newton's Principia, by I. Bernard Cohen*, with contributions by Michael Nauenberg, and George E. Smith, University of California Press, 1999, pp. 110–111. It is this: The forces considered in the second law are impulses, while the forces concerned in Newton's discussion of the first law are continuous. Thus, Cohen concluded that the first law is not a special case of the second, since different kinds of force are involved. While there is certainly a difference between the two kinds of forces, there seems to be no explicit mention of the difference in Newton's description of the two laws. If there is such an ambiguity in the occurrences of "force" in the First and Second Laws, why wasn't that difference disambiguated explicitly?

[4]*Matter and Motion*, 1876, recent edition Dover Press.

[5]Newton thought that there could be various forces acting on a body when he said "Corollary I" A body, acted on by two forces simultaneously, will describe the diagonal of a parallelogram in the same time as it would describe the sides by those forces separately", *Sir Isaac Newton's Mathematical Principles Of Natural Philosophy And His System Of The World*, Tr. F. Cajori, University of California Press, 1947, p. 14.

[6]Thus we are in full agreement with J. Earman And J. Robert's construal of Newton's Law of Gravitation as holding even in the presence of other non-gravitational forces, in their paper *Ceteris Paribus*: There Is No Problem of Provisos", Synthese, 118, 439–478, 1999, fn.14. This is in sharp contrast with N. Cartwright's version of the Newtonian Law of Gravitation, *How the Laws of Physics Lie,* Oxford University press, 1983, according to which the law holds when only gravitational forces are acting on a body. That is hardly ever the case, because other forces are usually present. As Earman and Roberts put it, "the law is irrelevant to real world situations". This is a good example of strange consequences sometimes following from strange assumptions.

According to the Second Law then, there are various forces that singly, and simultaneously have associated accelerations. It is this fact about the various applications of the Second Law that supports the idea that it is neither true nor false, but *schematic*.

This needs some explanation. We said that there are various applications or uses of the Second Law. If we are concerned with the motion of the planets, then the force function between the bodies a and b is the gravitational force FG i.e.

$$FG = (m_a \times m_b) \times G)/r^2,$$

where the gravitational force is directed along the line connecting the center of the two bodies, and is proportional to the product of the masses of the bodies, divided by the square of the distance between them. Replacing "F" in the Second Law with this particular function FG yields *Newton's Theory of Gravitation*. It is obtained as a special case or application of the Second Law. In other cases, such as Hooke's Law for a stretched spring, "F" is replaced in the Second Law by the function FH = -x, where x is the displacement when the spring is stretched from its equilibrium position by x units of length.

Newton thought that in these other cases, there would be other forces, and other laws: for optics, special forces would be needed to explain the diffraction of beams of light, and other special forces would be needed for chemical reactions, electrical, magnetic phenomena, and so forth. The idea was that for various problems, various forces would be used.[7] One force function cannot do the work for all the cases that the theory covers.

We have therefore situations in which there are different applications of the Second Law that involve different replacements of the letter "F". The interesting question is how to understand this apparently simple idea of replacement. How are we to think of this "F" that is replaced by various specific functions? In other words, what kind of expression is the "F" in "F = m x a"?

The answer, I suggest, is that "F" in the expression of Newton's Second Law is a schematic letter – a *function schematic letter*. Schematic letters, as we noted, come in various types: schematic sentential letters, schematic predicate letters and schematic function letters (for example).

The idea is roughly that wherever in some expression there are occurrences of (say) a type of schematic sentence letter, then all the occurrences of that schematic letter can be replaced by some particular sentence, predicate, or function. So for

---

[7]The one interesting objection to the Second law as specifying a condition for each of various forces that is known to me, was the proposal of the Eighteenth Century Croation physicist Roger J. Boscovich, *Theoria philosophiae naturalis*, Latin-English translation of the Venetian edition 1763. tr. by J.M. Child, Open Court Publishing Company, Chicago 1922. He declared that there was only one force, which varied with the distance between bodies, being sometimes repulsive and sometimes attractive, depending on the distance, but repulsive without limit when the bodies got closer beyond limit. J.C. Maxwell has an extended but dismissive discussion of Boscovich's view in his article on the Atom reprinted in his *Collected Papers*, Dover Press.

example if in the expression "S v P", where S is a schematic sentence letter, and P is a particular sentence, the result of the replacement could be "Q ∨ P", where Q is some particular sentence that is not schematic. Similar conditions hold for schematic predicate and function letters.

What we suggest then, is that the "F" in "F = ma" is a *schematic function letter* that can be replaced by terms for particular functions.[8] These schematic letters of various kinds, are thought of as place-holders, that can be replaced by particular sentences, relations, or functions.

There are two observations and one caveat prompted by the schematic character of Newton's Second Law.

(1) The first is that the special cases, the applications of a schematic expression, do not follow logically from it. So "FG = m x a", *Newton's Law of Gravitation*, which results from the replacement of "F" in the expression of the Second Law by the specific gravitational force function FG, doesn't follow from Newton's Theory of mechanics, and neither does Hooke's Law, FH = m x a. Clearly then, Newton's Law of Gravitation is not a consequence of his theory of mechanics, although it is a law that is one of the great successful applications of the Newtonian Theory of Mechanics.[9]
(2) The second observation is that the schematic Second Law is neither true nor false. While "F = m x a" has an occurrence of a schematic letter "F", one has to be careful. The fact that "F" is a schematic letter would not guarantee that the equation failed to have a truth-value. The reason is that even if an expression has a schematic letter as a part, the expression could still have a truth-value. For a trivial example, consider the expression "S v (P ∨ ¬P)", where "S" is a schematic sentence letter, and P is some definite sentence (say "Snow is white."). The schematic sentence will be true because "Snow is white" is true, and that implies S v (P ∨ ¬P).

However, the case for the equation "F = m x a" with schematic "F" is different, and requires a supporting argument. Here is one: We assume that if there is some semantic value assigned to the equation, it will have to assign semantic values to the items on both sides of the equation (either the same value, or different ones). But that is impossible, because no semantic values are assigned to schematic letters. They are place-holders for items that have semantic value, but they themselves have no semantic value. Consequently, the Second Law has no truth-value. The conclusion then for Newtonian theories is this: Newton's Theory of Mechanics is schematic, because "F" in the Second Law is a schematic letter.

---

[8]The use of schematic letters of sentences, predicates, and functions rather than variables of the appropriate type, allows them to be replaced by specific sentences, predicates and relations. This is not permitted if for example, "F" in our formulation of Newton's Second Law were required to be a variable ranging over functions.

[9]This is not the only example of an important physical theory that has successful applications that are not among its logical consequences. Cf. S. Morgenbesser and A. Koslow, "Theories and their Worth", *Journal of Philosophy* 2012, pp. 616–647.

Before we consider the status of Newton's Theory of Gravitation it is important to consider some remarkably prescient observations of Thomas Kuhn about how Newton's Second law gets to have empirical content, which it wouldn't otherwise have. In brief, it is by considering various *applications* of that law. It is worth considering his observation in full because he seems to have come to the conclusion that the Second Law is *schematic* (his term), when one considers the scientific practice of associating various exemplars of forces for various problems in order to provide an empirical content to the law, and without which it would have little empirical content.

> After the student has done many problems, he may gain only added facility by solving more. But at the start and for some time after, doing problems is learning consequential things about nature. In the absence of such exemplars, the laws and theories he has previously learned would have little empirical content.
>
> To indicate what I have in mind I revert briefly to symbolic generalizations. One widely shared example is Newton's Second Law of Motion, generally written as $f = ma$. ...
>
> That expression proves on examination to be a law-sketch or law-schema. As the student or the practicing scientist moves from one problem situation to the next, the symbolic generalization to which such manipulations apply changes. For the case of free fall, $f = ma$ becomes $mg = d^2s/dt^2$ ; for the simple pendulum it is transformed to $mg = -ml\ d^2\theta/dt^2$; for a pair of interacting harmonic oscillators it becomes two equations, the first of which may be written $m_1 d^2 s_1/dt^2 + k_1 s_1 = k_2 (s_2 - s_1 + d)$; and for more complex situations, such as the gyroscope, it takes still other forms, the family resemblance of which to $f = ma$ is still harder to discover. Yet, while learning to identify forces, masses, and accelerations in a variety of physical situations not previously encountered, the student has also learned to design the appropriate version of $f = ma$ through which to relate them, often a version of which he has encountered no literal equivalent before. How has he learned to do this?"[10]

The agreement between Kuhn's observations about Newton's Second Law, and my own conclusion are very welcome. The views are very close. Kuhn wrote about "exemplars"; I called then "applications". He used different equations; I wrote of different force-functions that were substituted in F = ma. The equations are just what you get when you unpack "F = ma" for the different forces. He said that without the exemplars, $f = ma$ would have little empirical content, whereas I think that that equation has no truth value as it stands. Kuhn thought that the exemplars he mentioned have a family resemblance. I don't agree on that relatively minor difference. There is, as far as I can see, no family resemblance say between the gravitational force FG, and the Hook force, nor is there an evident resemblance that holds between Kuhn's examples of exemplars. The various applications of the schematic Second Law are all subsumed under it, but they needn't have, nor do they have a family resemblance.

---

[10]T.S. Kuhn, *The Structure of Scientific Revolutions*, Second Edition, Enlarged, volume 11 Number 2, The University of Chicago Press, Volume 11, Number 2, pp. 188–189. Thanks to an anonymous referee for calling my attention to this reference to Tom Kuhn's comments on Newton's Second Law.

The use of “law-schema” is an amazing agreement. On my usage, the schema has no truth value -but the subsumed items do. Kuhn thought that the schema has little empirical content. Nevertheless, the two views are fairly close -especially in their appeal to scientific practice for their cogency. We turn next to a consideration of Newton’s theory of Gravitation

Newton’s theory of Gravitation (essentially the subject of Book III of his *Principia*) however, is the theory that results by replacing the schematic letter “F” in the Second Law by the specific gravitational force FG. It refers to a specific function, and is not a schematic letter. Newton’s Theory of Gravitation therefore is not a schematic theory. Moreover, Newton’s Law of Gravitational force is not a deductive consequence of his theory of mechanics. This then is a case of a law that does not follow deductively from its theoretical scenario. – Newton’s theory of Classical Mechanics.

Generally, in the case when a law L has a schematic theory as a theoretical scenario, the associated magnitude vector space for L is constructed exactly as in the cases previously described. The magnitudes for the associated magnitude vector space come from two sources: one consists of the magnitudes provided by the schematic theory, and the other is provided by the magnitude that was used to replace the schematic letter in order to get L. For example, in the case of Newton’s Law of Gravitation, the magnitudes are provided respectively from the schematic theory of mechanics (i.e. mass, velocity, and acceleration), and the magnitude of gravitational force (FG) that was used to replace the schematic letter “F” in the Second Law).

### *12.1.2 The Lagrangian Theory of Mechanics*

This theory is usually understood to be more general than the Newtonian theory of mechanics, with certain advantages for obtaining solutions to some problems in mechanics that would be extremely difficult using the Newtonian theory of mechanics. It is well known that by making various substitutions for its schematic letters, one obtains (1) Newton’s Law of Inertia, as well as (2) the conservation of energy, (3) the conservation of linear momentum and (4) the conservation of angular momentum. Thus, several important laws of mechanics are subsumed under Lagrange’s theory of mechanics, and none of them are logical consequences of it. We look first at how (1) is obtained, and indicate how the cases go for (2) – (4).

The *Lagrangian Function* of a finite system of particles is a function of q1, q2, . . . , qn, which are generalized positions, and q1’, q2’., . . . , qn’ which are the generalized velocities of the of the particles of the system (we use the prime notation to indicate differentiation with respect to time), and is given by

$$\textbf{(LF)}\ L(T - U),$$

where T is the total kinetic energy of the system (1/2)Σ mqi' [2], and U is the potential energy U(qi, . . ., qn), – a certain function of the coordinates of the particles.

In the applications of this theory to various different systems of particles, the expressions for T and U will generally be replaced by different functions that are appropriate to the systems under consideration. For that reason, we think that the Lagrangean theory is schematic. The argument in this case is essentially the same as the one we used to support the belief that Newton's Theory of Mechanics is schematic.

Here is a sketch of how the subsumption of Newton's first law of intertia under Lagrangian Theory can be shown:

Newton's First Law first law is stated this way[11] :

> (**LL**) "In an inertial frame of reference, any free motion takes place with a velocity which is constant in both magnitude and direction."

An *inertial frame* is defined (following Landau and Lifshitz) as a frame of reference in which space is homogeneous, and isotropic, and time is homogeneous. Space being homogeneous means that various positions in space are mechanically equivalent. Isotropy requires that different orientations in space are mechanically equivalent. Both conditions are spelled out in terms of conditions (sometimes called symmetry conditions) that are placed upon the Lagrangian in an inertial frame of reference and a free motion is one with no external forces acting on it. Thus, Landau and Lifshitz say:

> The homogeneity of space and time implies that the Lagrangian cannot contain explicitly either the radius vector **r** or the time $t$, i.e., $L$ must be a function of the velocity only. Since space is isotropic, the Lagrangian must also be independent of the direction of **v**, and is therefore is a function only of its magnitude, i.e. $\mathbf{v}^2 = v^2 : L = L(v^2)$.

There are well known ways of deriving the Lagrangian Equations of motion (**LE**), using this particuar Lagrangian. i.e.

$$\text{(LE)}\ \ d/dt(\partial L/\partial \mathbf{v}) = \partial L/\partial \mathbf{r}.)$$

Now (∂L/∂r) is 0 because r does not occur in the Lagrangian, so that d/dt (∂L/∂v) is 0, and consequently, ∂L/∂v is a constant. But, as Landau and Lifshitz noted, it is a function of the velocity only. Consequently, **v** is constant. So we have that Newton's first law of motion is subsumed under Lagrangian mechanics.

There are other important generalizations that are also subsumed under Lagrangian schematic theory. I.e.:

---

[11]This sketch follows the elegant presentation of that result including some other familiar conservation laws given in L. D. Landau and E. M. Lifshitz (*Mechanics, volume 1 of Course of Theoretical Physics*, Tr. By J.B. Sykes and J.S. Bell, Pergamon Press, Addison-Wesley Publishing Co, Reading Mass, 1969), Chapters I and II. A more mathematical presentation of these results together with more advanced mathematical results can be found in V.I. Arnold's *Mathematical Methods of Classical Mechanics, Second Edition,* 1989, Chapters 3 and 4.

(1) Assuming the homogeneity of time, we have the conservation of energy for closed systems.
(2) Assuming the homogeneity of space, we have that the sum of all forces in a closed system is zero. (Newton's Third Law)
(3) Assuming the isotropy of space, we have the conservation of angular and of linear momentum.

Note that these examples of subsumption are laws, but not every generalization that is subsumed is a law.

### *12.1.3 The Hamiltonian Theory*

We have already alluded to the Hamiltonian theory of mechanics in Chap. 10. Those comments, together with some ideas proposed by the Mathematician H. Whitney were used to motivate the concept of a Measurement Vector Space (**MVS**) of any theory that is expressed with the aid of magnitudes. The elements of such a vector space were then taken to be the physical possibilities provided by that theory. The Hamiltonian Theory of Mechanics we believe is also a schematic theory

Recall that when we mentioned the Hamiltonian in motivating our notion of theoretical possibilities, we took the Hamiltonian of a system of particles to be given by a function H(p1, ... , pn, q1, ... , qn, t) for any n-bodied system, where the pi correspond to what are called the generalized momenta of each of the n particles, and the corresponding qi are the generalized positions. These are usually regarded as the basic magnitudes of the theory.

Contemporary expositions of Hamiltonian mechanics usually rest on a very sophisticated branch of mathematics (Simplectic Geometry), and usually cover a wide range of physical as well as mathematical theories. For the present purpose of emphasizing the schematic character of the theory, we restrict our remarks to the Hamiltonian theory of classical mechanics. [12]

In the case of classical mechanics, the Hamiltonian equations are given by:

$$\textbf{(HE)} \quad \begin{aligned} \mathrm{qi}' &= (\partial H/\partial pi) \\ \mathrm{pi}' &= -(\partial H/\partial qi). \end{aligned}$$

[12]For a very lucid and thorough mathematical exposition, V.I. Arnold, *Mathematical Methods of Classical Mechanics,* Second edition, Springer 2010, Chapters 7, 8, and 9. For an illuminating discussion of the different ways in which radically different Hamiltonian functions can figure in a variety of physical and mathematical theories, cf. Terence Tao, "Hamiltonians", in *The Princeton Companion to Mathematics,* Timothy Gowers, (ed.), J Barrow-Green and I Leader (associate eds.), 2008, III.35 pp. 215–216.

Now consider the special case when the Hamiltonian function does not depend explicitly on the time t. For that special class of cases, the energy of the system is conserved. The reason is that

$$\mathrm{dH/dt} = \Sigma i\,[(\mathrm{dH/dpi})\ (\mathrm{dpi/dt}) + (\partial H/\partial qi)(\mathrm{dqi/dt})] + .\partial H/\partial t.$$

Using **(HE),** we see that $\Sigma$i [(dH/dpi) (dpi/dt) + ($\partial$H/$\partial$qi)(dqi/dt)] = 0, so that dH/dt = $\partial$H/$\partial$t. However, in this special case, where t is not explicit in the Hamiltonian, it follows that $\partial$H/$\partial$t = 0. Consequently, dH/dt = 0. The conclusion therefore is that the Hamiltonian i.e. the energy of the system, is conserved.

It is well known that with judicious choice of different Hamiltonian functions.in **(HE)**, conservation laws like (3) and (4) are also subsumed under Hamiltonian theory.

The schematic status of Newton's theory of Mechanics, the Lagrangian theory, and Hamiltonian theory of mechanics are schematic where "F", "L", and "H" are understood to be schematic function letters. It is not true that one force function, one Lagrangean, and one Hamiltonian is appropriate for all those cases that these theories cover.

In summary then, Newton's Theory of Mechanics, Lagrangian, and Hamiltonian Mechanics are schematic theories. Each subsumes some familiar laws, but none of those laws are deductive consequences of the respective schematic theories, nor do those theories explain anything, if factivity is required of explanations. Nevertheless each of these schematic theories are regarded as powerful and prized. That high regard is a measure of all of the successes of those theories i.e. the results that are subsumed under them.

There are two other features which contribute to the high esteem that these schematic theories enjoy: one is that these theories can explain∗ (as defined in Chap. 11) those features which are explained by those generalizations they subsume. The other is that the generalizations that they explain∗ are not only true, but *non-accidental*. In the next Sect. (2), we will turn to a discussion of these two features.

However, before we turn to that task, we would like to explore briefly, some special features of a specific theory that, in its original presentation, was considered to be schematic by its author. It is the well- known classical axiomatization of probability by A. Kolmogorov. It appears to be unlike the other schematic theories we have described, because it also appears to be a modal theory. It is a schematic theory that subsumes various important applications under it, some of them mathematical and others physical, but none of which appear to be laws. Nevertheless, it is, we believe, of special interest since it is also, as we shall see, an example of a modal theory.

### 12.1.4 Kolmogorov's Schematic Axiomatization of Probability

We think that A. Kolmogorov's well known theory of probability was proposed by him as what we have called a schematic theory. This runs counter to many, perhaps all current presentations of that theory, but it was, as we shall see below, a schematic version that Kolmogorov intended, – inspired, as he admitted, by Hilbert's work in Geometry. The usual contemporary accounts of probability are described as distinct theories, assigning different meanings to "probability": frequency, geometric, measure-theoretic, dispositional, and subjective, to name a few. Rather than appeal to a diversity of meanings, the schematic theory appeals instead to a diversity of applications, each obtained by replacing the schematic letter for "probability" by different replacements. One replacement yields Lebesgue Measure theory, other replacements yield geometric probability, and some other replacements may yield a frequency theory of probability

I will set to one side the host of qualitative non-quantitative theories that diverge from Kolmogorov's theory in various interesting ways. The impression given by the current plethora of theories is that "probability" is ambiguous, with proponents in sharp disagreement about which interpretation, if any, is correct.

The idea that Kolmogorov's theory is schematic throws some light on how to understand these different "interpretations". If probability theory is a schematic theory, there are a variety of applications of the theory. One doesn't have to choose which of them, if any, is the correct one. One application does not rule out another.

They are all applications, each subsumed under Kolmogorov's schematic probability theory; each obtained by a replacement of the specific schematic function letter "P". For Kolmogorov, and for us, the variety is not a fault of the theory, it is a virtue.

We turn to a consideration of whether Kolmogorov thought of probability theory as schematic.

### 12.1.5 The Kolmogorov Axiomatization of Probability as a Schematic Modal Theory

This section is devoted to three claims. (i) That Kolmogorov thought he had produced a much needed axiomatization of probability theory and that it was schematic. (ii) That we think that his probability theory provides a concept of probabilistic-truth which has some intrinsic interest, even though it fails to satisfy the Tarski condition for a truth operator, and (iii) The notion of probabilistic-truth can be shown to be a modal notion, so that probability theory is a powerful schematic theory that is modal.

(i) The impetus to axiomatize probability theory seems to have been Hilbert's 1900 list of open mathematical problems, one of which was to axiomatize physics as well as Hilbert's slightly earlier publication *Foundations of Geometry* (1899). Kolmogorov said that

> ... the concept of a *field of probabilties* is defined as a system of sets which satisfies certain conditions. What the elements of this set represent is of no importance in the purely mathematical development of the theory of probability (cf. the introduction of basic geometric concepts in the *Foundations of Geometry* by Hilbert, of the definitions of groups, rings, and fields in abstract algebra. (1).

This idea is then immediately followed by a marvelous passage that we believe supports the view that the resultant theory is schematic:

> Every axiomatic (abstract) theory admits, as is well known, of an unlimited number of concrete interpretations besides those from which it was derived. Thus we find applications in fields of science which have no relation to the concepts of random event and of probability in the precise meaning of these words. (1).[13]

Here he indicates that any axiomatization has infinitely many interpretations. That is, there are infinitely many distinct fields of probabilities with distinct elements, and distinct probability functions defined over them. This echoes exactly the way that Hilbert spoke of axiomatizations in his correspondence with Frege that we described in Chap. 9.

What is even more noteworthy, and striking, is that he thought that these various applications of probability theory could include mathematical examples as well as physical ones. The prime mathematical example appears to be H. Lebesgue's theory of measure and integration, and he also mentions a version of R. von Mises frequency interpretation which may be what he had in mind when he said that there were applications in fields of science. Although Kolmogorov thought that the von Mises frequency theory was an attempt to define probability in terms of more basic concepts, nevertheless, the end result would be, an empirical development of the theory.

We should note that the influence of Hilbert's view of a successful axiomatization is reinforced by two of Kolmogorov's observations. The first is that his axioms form a consistent set -one of the prime considerations emphasized by Hilbert. The (relative) consistency was provided by a simple model (p. 2). The second observation was expressed with the help of an unusual notion of "completeness". He notes that

> Our system of axioms is not, however, *complete*, for in various problems in the theory of probability different fields of probability have to be examined (p. 3).

Kolmogorov's unusual use of "completeness" is not standard, but we think that his explanation conveys the idea that his axiomatization of the theory of probability is

---

[13]The two quotations are from A. N. Kolgmogorov, Foundations of the Theory of Probability, second English Edition, translation N Morrison, added bibliography by A.T. Bharucha-Reid, Chelsea Ppublishing Company, New York, 1956. p. 1., Original German, 1933.

schematic, because, as he noted, there are many different probability functions that are used in various different applications of the theory.

We now turn to a discussion of an operator, which we, for a want of a better term, will call *probabilistic-truth.* It is definable within Kolmogorov's theory, and seems to mirror a truth operator – up to a point. It behaves, as we shall see, just like a truth operator would, with respect all the logical operators, except for classical negation, where it only goes half the way one would expect; it fails to satisfy the Tarski biconditional.

Nevertheless this divergence can be explained we think by noting that there is a modal operator closely connected to the concept of probabilistic truth. We turn first to the behavior of probabilistic-truth with respect to the logical operators.

### 12.1.6 Probabilistic-Truth and the Logical Operators

Consider the sentential operator[14] PT, which to each A, assigns the sentence p(A) = 1. i.e. "'A' is probabilistically-true", and "p(A) is 0" is to be read as "'A'" is probabilistically-false. The following are then easy consequences of Kolmogorov's theory:

1. If A1, A2, ..., An implies B, and all the p(Ai)s are 1, then so too is p(B).
2. If p(A) is 1, then p(¬A) is 0, (if negation is classical), and conversely.
3. p(A ∧ B) = 1 if and only if p(A) is 1 and p(B) is 1.
4. p(A ∨ B) is 1 if either p(A) is 1, or p(B) is 1.
5. p(A ∨ B) is 0 if and only if p(A) and p(B) are both 0.
6. If p(A) is 0, then p(A ∨ B) = p(B).
7. If p(A → B) is 1, and p(A) is 1, then p(B) is 1. (for the material conditional).
8. If p(A → B) is 0, then p(A) is 1, and p(B) is 0.

There is the interesting fact that that probabilistic-truth falls short of satisfying the Tarski T-schema for truth- i.e that for all A, T(A) implies and is implied by A – i.e. T(A) ⇔ A, where "T" is a truth operator.

To see that, we consider a condition that the Tarski schema implies, and show that probabilistic-truth fails to satisfy it.

We know that PT(A) and PT(¬A) together are inconsistent, since their conjunction implies that p(A) is equal to 0 and also to 1. Therefore, PT(A) ⇒ ¬PT(¬A).

[14]We have noted above that Kolmogorov's formulation of probability is schematic. There are formulations in which probabilities are assigned to events and other formulations where the assignment is to sentences. That possibility is permissible for schematic theories. Nevertheless the two versions are not equivalent. That important but overlooked point is made clearly by A. Hajek and C. Hitchcock in their introduction to *The Oxford Handbook of Probability and Philosophy*, Oxford University Press, 2016, p. 20. In the following section, we are using a formulation in which probabilities are assigned to sentences rather than events, for the simple reason that here we are interested in probabalistic-*truth.*

The converse $\neg PT(\neg A) \Rightarrow PT(A)$ fails. If it didn't, then the equivalence $PT(A) \Leftrightarrow \neg PT(\neg A)$ holds. Then p(A) is 1 $\Leftrightarrow$ p(A) is not 0. That is, for every A, the probability is either 0 or 1. Such probability functions are called *trivial*. Thus we have the simple result that

For every non-trivial probability function, the equivalence $PT(A) \Leftrightarrow \neg PT(\neg A)$ fails for every A.

The failure of PT to satisfy the Tarski Truth schema ($T(A) \Leftrightarrow A$) follows.

### *12.1.7 The Modal Character of Kolmogorov's Theory of Probability*

Here we will make use of the notion of a Gentzen structural modal[15] that was introduced and explained in Chap. 10. Let S be a non-empty set of sentences, with a Gentzen implication relation $\Rightarrow$, on it. We remind the reader that $\varphi$ is a *Gentzen modal operator* on S if and only if it is an operator, on S[16] with values in S, such that the following two conditions hold:

(I) If for any sentences A1, A2, ..., An, and B, in S, such that if A1, A2, ..., An $\Rightarrow$B, then $\varphi$(A1), $\varphi$(A2), ..., $\varphi$(An) $\Rightarrow$ $\varphi$(B), and
(II) For some A and B in S, $\varphi(A \vee B)$ fails to imply [$\varphi$(A) v $\varphi$(B)].

We now define an operator T$*$ in terms of the notion of probabilistic-truth as follows:

For any A in S, let T$*$(A) be A, if $p(A) = 1$, and a contradiction, $\kappa$, otherwise.

The claim we wish to make is that T$*$ is a Gentzen modal operator. So the task at hand is to verify that conditions (I) and (II) are satisfied. In this case, then we have to look at the corresponding:

(I′) If A1, A2, ..., An $\Rightarrow$B, then T$*$(A1), T$*$(A2), ..., T$*$(An) $\Rightarrow$ T$*$(B). and
(II′) For some Ao and Bo in S, T$*$(Ao $\vee$ Bo) fails to imply [T$*$(Ao) v T$*$(Bo)].

Consider (I′). Assume that A1, A2, ..., An $\Rightarrow$ B . Suppose that T$*$(A1), T$*$(A2), ..., T$*$(An) and that for some j, p(Aj) is not equal to 1. Then T$*$(Aj) is $\kappa$, a contradiction. Therefore T$*$(A1), T$*$(A2), ..., T$*$(An) implies everything in S, and in particular

[15]This kind of modal operator as we mentioned in Chap. 11, fn. 2, is studied extensively in A. Koslow, *A Structuralist Theory of Logic*, Cambridge University Press, Cambridge 1992, Part IV, pp. 239–371, and "The implicational Nature of Logic: A Structuralist Account", in *European Review of Philosophy, The Nature of Logic*, volume 4, edited by A. Varzi, 1999, CSLI Publications, Stanford, pp. 111–155, where it was simply called a modal.

[16]We assume here that S is closed under the classical logical operators.

T∗(A1), T∗(A2), ..., T∗(An) ⇒ T∗(B). On the other hand, if p(Ai) is 1, for all the Ai, it follows that p(B) is 1, so that T∗(B) is B. Moreover, since all the p(Ai) are 1, it also follows that for every i, T∗(Ai) is Ai. Consequently since A1, A2, ..., An ⇒B, we also have that T∗(A1), T∗(A2), ..., T∗(An) ⇒ T∗(B). So (I) is satisfied.

Consider (II′). Here we need the assumption that the probability function on S is not trivial. Consequently there is some Ao in S, such that P(Ao) is neither 0, nor 1. It follows also from the axioms of probability that the probability of ¬Ao is also neither 0 nor 1. Does T∗(Ao v ¬Ao) imply T∗(Ao) v T∗(¬Ao)? It doesn't. T∗(Ao v ¬Ao) is Ao v ¬Ao, because the probability of Ao v ¬Ao is 1. And T∗(Ao) and T∗(¬Ao) are both equal to κ, a contradiction. Since (Ao v ¬Ao) doesnt' imply a contradiction, condition (II′) is satisfied.

Thus the operator T∗ defined in terms of probable-truth is a Gentzen structural modal. It is not a widely known modal operator, but some of its features are easily proved. We list some of them below:

(i) T∗(A) ⇒ A, for all A. There are two cases: (1) if p(A) = 1, then T∗(A) is A, and A ⇒ A. If p(A) differs from 1, then T∗(A) is a contradiction, and so it implies A.
(ii) A ⇒ T∗(A) ***fails*** to hold for some A. It fails for some Ao for which p(Ao) differs from 1 and from 0. There is such an Ao because the probability is assumed to be non-trivial. In that case, T∗(Ao) is a contradiction, and so p(Ao) is 0.
(iii) T∗(A) ⇒ ¬T∗(¬A), for all A. This holds because the conjunction T∗(A) ∧ T∗(¬A), implies the contradiction (A ∧¬A) by (i). Therefore T∗(A) ⇒ ¬T∗(¬A), for all A.
(iv) ¬T∗(¬A) ⇒ T∗(A) ***fails*** for some A.[17] Since the probability function is assumed to be non-trivial, there is an Ao such that the probability p(Ao) is neither 0 nor 1. And it follows also that the probability of ¬Ao is neither 0 nor 1. Consequently, T∗(Ao) is a contradiction κ, and ¬T∗(¬A) is the negation of the contradiction κ. And that is impossible.

After this brief excursus about schematic probability theory, we return to the consideration of the explanatory power of schematic theories. We want to show that if a schematic theory explains any contingent generalization, then that generalization is non-accidental. This parallels the similar conclusion for theories shown in Chap. 7.

[17]It is interesting to note that this modal implies its dual but not conversely. That is, for any A, T∗(A) ⇒ ¬T∗(¬A), but not conversely (unless the probability function p is trivial).

## 12.2 Subsumtive Explanation∗, Truth, and Non-accidentality

We believe that schematic theories tell us something important about those generalizations that are subsumably explained by them:

**(TNA)** The generalizations that are explained∗ by some schematic theory T#, are true generalizations and are not accidental.

The proof of this result is not deep. Let T# be a schematic theory, and suppose that T# sumsumtively explains∗ some contingent generalization E. Then E is a true non-accidental generalization.

The reason is that if E is explained∗ by T#, then there is some statement S that is subsumed under T∗ which explains E. (note here that it is "explains", and not "explains∗" that is required, and it is not required that S is a law). Since S explains E, we know from the result of Chap. 7), that E is true by the factivity of "explains". It was also shown, using our mini theory (**LAG**) in Chap. 7, that if there is an explanation of a generalization E, then E is not accidental. In summary then:

(1) If there is an explanation of a contingent generalization E, then E is a true non-accidental generalization. This result of Chap. 7, implies
(2) If there is a theory that explains a contingent generalization E, then E is a true non-accidental generalization.

And now there is this generalization for schematic theories:

(3) If there is a schematic theory that explains∗(explanatorily subsumes) a contingent generalization E, then E is a true, non-accidental generalization.

## 12.3 A Connection Between Explanation and Explanation∗

Recall (Chap. 9) that a schematic theory U subsumtively explains (explains∗) A, if and only if there is some non-schematic B that is subsumed under U, and B explains A. In that case, there is a nice connection between between "explains∗", and "explains".

Suppose that U is not a schematic theory, and it explains A. Then replace the sentences, and relations in U by schematic letters of the appropriate kind. Thus, for example, if U is the theory (x) Fo x → Go x), with the specific predicates "Fo x" and "Go x", replace them by the schematic letters "F∗ x" and "G∗ x" to obtain the schematic statement U∗: (x) F∗ x → Go∗x). Notice now that U is subsumed under U∗, and U explains A. Therefore the schematic U∗ subsumtively explains (explains∗) A. It is easy to see how to proceed generally. Therefore we know that there is a connection between explanations and explanations∗s:

**(EE∗)** If a non-schematic theory U explains some A, then there is a schematic theory U∗ (the schematization of U) that explains∗ A.

In other words, any A that is explained by a non-schematic U is also explained∗ by a schematic U∗.

## 12.4 Generality Redux. The Generality of Theories Compared to the Laws That They Explain

There is, I believe, a kind of generality to schematic theories. It is not the kind of generality that a universally quantified sentence (x)Hx provides for its deductive instances H(a), H(b), H(c). ... Instead it is the kind of generality that a schematic theory U has in its gathering up all the subsumed instances, each of which is the uniform result of the substitution of appropriate items for the various schematic letters in the theory. In sum, schematic theories have a *subsumtive generality.*

There is of course the important difference between the two cases: all the instances H(a), H(b), Hc),..., of the universally quantified H(x) are deductively subsumed under (x)H(x), whereas the instances of a schematic theory U are not deductively subsumed under U; they are schematically subsumed under U. We think that any schematic theory also has greater generality than any item that is schematically subsumed under it. That is because what schematic theories subsume are just special cases of that theory obtained by appropriate substitutions.

Beyond whatever generality that laws and theories may have on their own, there is the different question of whether theories have greater generality than those laws that they explain, – a different matter altogether.

Having recognized that theories could come in at least two kinds, schematic and non-schematic, we described two kinds of explanation that these theories can provide: explanation∗ (subsumtive explanation) for schematic theories and explanation (simpliciter) for those theories that are not schematic. **(EE∗)** is one way of seeing how the concepts of explanation and explanation∗ are related.

We turn now from a discussion of those two kinds of explanations, to a brief consideration of whether theories that explain laws have at least the generality of those laws that they explain. It is a topic that has had very little discussion[18]

I do not know of any satisfactory definition of the relation of "is at least as general as" that would make it evident that schematic and non-schematic theories are at least as general than those laws that they explain. However there are a few observations which I think may be of some interest, even though they do not go as far as one would hope.

[18]Cf. the accounts of E. Nagel and R.B. Braithwaite in Chap. 8.

We want to tentatively propose a mini-theory (**GTL**) that has three assumptions so far, that specify some conditions connecting theories, laws, explanations and the concept of generality.

To begin, note that schematic theories are at least as general as any A, law or not – that is subsumed under it. Let "X >> >Y" stand for "X is at least as general as Y". So, for any schematic theory T∗, and any A that is subsumed under it, we shall assume that

(1) $T^* > > > A$.

The reason is that A is obtained as a special case of T∗ by replacing the schematic letters of T∗ by appropriate specific examples. It seems reasonable therefore to endorse (1).

In the last section we proved that if V is a non-schematic theory, and V∗ is a schematized theory associated with V, then for any non-schematic theory V and its schematization V∗, V is subsumed under V∗. So we shall assume that for such V and V∗, that

(2) $V^* > > > V$.

We will now relax the strategy of looking for a definition of "is at least as general as", in favor of finding some plausible conditions about generality. We might make some progress if we can find a plausible condition that makes use of the concept of explanation. Here is one proposal of that sort that is worth considering:

Suppose that a theory U is at least as general as a theory V, and that V provides an explanation for some A (i.e. Exp[V; A]). I.e. A is explained by some theory V, and U is at least as general as the theory V. It seems initially plausible then, under these conditions, that U is at least as general as A. So let's provisionally propose that

(3) If U >>> V, and there is an explanation that V provides for A (i.e. Exp[V; A]), then they together imply that U is at least as general as A. i.e.U >>> A.

This is a condition relating the condition of greater generality with the concept of explanation. We can now derive from (2) and (3) the following results for the case when U is schematic, and for the case when U is not schematic.

(i) U is a schematic theory. Let U be Newton's Theory of Classical Mechanics, – a schematic theory. And let V be Newton's Theory of Gravitation, -a non-schematic theory that is subsumed under it. Let K be Kepler's first law that all the planets have elliptic orbits. There is an explanation that the Theory of Gravitation provides for Kepler's first law. Consequently, by (3) we have the nice result that
   Newton's Classical Theory of Mechanics is at least as general as Kepler's first law.
(ii) Let U be a non-schematic theory. Now U is at least as general as itself – i.e. U >>> U. Suppose further, that U is also an explanation of the law L. Again by (3), we have the result that U >>> L. In general,
   If any non-schematic theory U explains a law L, then U is at least as general as L.

Examples abound: Newton's theory of Gravitation is at least as general as Kepler's first law of planetary motion, or Galileo's law of falling bodies ....

Our mini-theory was tentatively proposed. There are many still unresolved questions that remain: (1) should the generality condition be strengthened? – that the explaining theory has greater generality than the law that is explained, and not, as we have it, the weaker requirement that the theory should be at least as general as the explained law. The strengthened condition would imply an asymmetry of explanation. Which is moot. (2) Is there a way of explicitly defining the generality relation that would justify or support the conditions of the mini-theory? Perhaps. These are still open questions as far as I can see.

# Bibliography


Aristotle. (1975). Posterior analytics (J. Barnes, Trans., pp. 78a23–78b15). Oxford University Press, PA, A13.

Armstrong, D. M. (1983). *What is a law of nature*. Cambridge: Cambridge University Press.

Armstrong, D. M. (1997). *A world of States of Affairs*. Cambridge/New York/Melbourne: Cambridge University Press.

Arnold, V. I. (1989). *Mathematical methods of classical mechanics* (2nd ed.). New York: Springer.

Azzouni, J. (2014, May 16). *A new characterization of scientific theories*. published online.

Bernays, P. (1998). *"Hilbert's significance for the philosophy of mathematics"*(1922) (Trans. by P. Moncosu, in Moncosu, From Brouwer to Hilbert. Oxford University Press.

Boscovich, R. J. (1922). *Theoria philosophiae naturalis*. Latin-English translation of the Venetian edition 1763 (J. M. Child, Trans.) Chicago: Open Court Publishing Company.

Braithwaite, R. B. (1923). *A disssertation on causality*, 1923, King's Fellowsip Dissertation, King's College Archives.

Braithwaite, R. B. (1953). *Scientific explanation*. New York: Cambridge University Press.

Bridgman, P. (1961). *The nature of thermodynamics*. New York: Harper Torchbooks. Harvard University Press, 1941.

Campbell, N. R. (1920). *Physics the elements*. Cambridge University Press.

Carroll, J. (2008). Nailed to Hume's Cross? In J. Hawthorne, T. Sider, & D. Zimmerman (Eds.), *Contemporary debates in metaphysics*. Oxford: Basil Blackwell.

Cartwright, N. (1983). *How the laws of physics lie*. New York: Oxford University press.

Cohen, I. B., Whitman, A., & Newton, I. (1999). *The principia a new translation*. Preceded by a Guide to Newton's Principia, by I. Bernard Cohen, with contributions by Michael Nauenberg, and George E. Smith. University of California Press

Corey, L. (2004). *David Hilbert and the axiomatization of physics: From Grundlagen der Geometrie to Grundlagen der Physik*. Berlin: Springer.

Courant, R. (1973). *Differential and integral calculus* (E. J. McShane, Trans.). Interscience Publishers, 1937v.I.

Cushing, J. (1998). *Philosophical concepts in physics*. Cambridge University Press.

Da Costa, N. C. A., & French, S. (2003). *Science and partial truth, a unitary approach to models and scientific reasoning*. Oxford: Oxford University Press.

Dijksterhuis E. J. (1961). *The mechanizaton of the world picture* (C. Dikshoorn, Trans.). Oxford University Press.

Dretske, F. (1977, June). Laws of nature. *Philosophy of Science*, *44*(2), 248–268.

Dudman, V. H. (1991). Interpretations of 'If' – sentences. In F. Jackson (Ed.), *Conditionals*. Oxford Readings in Philosophy, Oxford University Press.




A. Koslow, *Laws and Explanations: Theories and Modal Possibilities*, Synthese Library 410, https://doi.org/10.1007/978-3-030-18846-7

Frege, G. *Philosophical and mathematical correspondence* (G. Gabriel, H. Hermes, F. Kambartel, C. Thiel, A.Verart, Eds.), Abridged from German edition by B. McGuinness and trans. by H. Kaal.

Friedman, M. (1974). Explanation and scientific understanding. *Journal of Philosophy, 37*(I), 5–19.

Galavotti, M. C. (1991). *Notes on philosophy, probability and mathematics*. Bibliopolis.

Gentzen, G.(1933). Untersuchungen über das logische Schliessen. *Mathematische Zeitschrift, 39*, 176–210, 405–431; Reprinted (1969) as "Investigations into logical inference". North-Holland Publishing Co., pp. 68–131.

Giles, R. (1964). *Mathematical foundations of thermodynamics*. Oxford: Pergamon Press.

Glymour, C. (2013, April). Theoretical equivalence and the semantic view of theories. *Philosophy of Science, 80*, 286–297.

Grattan-Guinness, I. (1970). *The development of the foundations of mathematical analysis from Euler to Riemann*. MIT Press, Cambridge, MA.

Hajek, A., & Hitchcock, C. (2016). Introduction. In *The Oxford handbook of probability and philosophy*. Oxford: Oxford University Press.

Halvorson, H. (2003). *The semantic view, if plausible, is syntactic*. Oxford Handbook of Philosophy of Science (2015).

Halvorson, H. (2012, April). What scientific theories could not be. *Philosophy of Science, 79*, 183–206.

Hilbert, D., & Bernays, P. (1934). *Grunndlagen Der Mathematik*, vol. 1. Verlag von Julius Springer.

Hausdorff, F. *Mengenlehre*, First German edition 1914, Third English Edition *Set theory*. Chelsea Publishing Company, Trans. J. R. Aumann et al.

Hempel, C. G. (1966). *Philosophy of natural science*. Prentice-Hall.

Hempel, C. G. (1965). *Aspects of scientific explanation*. The Free Press: New York.

Johnson, W. E. (1964). *Logic*, Part I, 1921, Part II, 1922, Part III in 1924. Cambridge University Press. Reprinted by Dover Publications, Inc.

Kitcher, P. Explanatory unification. *Philosophy of Science, 48*, 507–531.

Kleiner, I. (1989). Evolution of the function concept: A brief survey. *College Mathematics Journal, 20*, 282–399.

Kolmogorov, A. N. (1956). *Foundations of the theory of probability*, second English Edition, translation N. Morrison, added bibliography by A. T. Bsrucha-Reid, Chelsea Publishing Company: New York. Original German 1933.

Koslow, A. *Explanation and modality*. Typescript.

Koslow, A. (2015). Laws, accidental generalities, and the Lotze uniformity condition. In *Conceptual clarifications*, Tributes to Patrick Suppes (1922–2014), Dov Gabbay. College Publications, CO.

Koslow, A. (2012). The explanation of laws: Some unfinished business. *The Journal of Philosophy, 109*(8/9), 479–502. Special Issue: Aspects of explanation, theory, and uncertainty: Essays in honor of Ernest Nagel, ed. by Bernard Berofsky and Isaac Levi.

Koslow, A. (2006). The representational inadequacy of Ramsey sentences. *Theoria, 72*(2), 100. Part 2.

Koslow, A. (2003). Laws, explanations and the reduction of possibilities. In H. Lillehammer & G. Rodriguez-Peryra (Eds.), *Real metaphysics essays in honour of D. H. Mellor* (pp. 169–183). London/New York: Routledge.

Koslow, A. (1992). *A stucturalist theory of logic*. Cambridge University Press.

Koslow, A. (1999). The implicational nature of logic. In A. C. Varzi (Ed.), *European review of philosophy, the nature of logic* (Vol. 4, pp. 111–155). Stanford: CSLI Publications.

Krantz, D. H., Duncan Luce, R., Suppes, P., & Tversky, A. (1971). *Foundations of measurement, volume I, Additive and polynomial representations*. New York: Academic Press.

Kuhn, T. S. (1990). *The structure of scientific revolutions* (Second Edition enlarged, vol. 11, Number 2). The University of Chicago Press.

Landau, L. D., & Lifshitz E. M. (1969) *Mechanics, volume 1 of course of theoretical physics* (J. B. Sykes, & J. S. Bell, Trans.). Reading: Pergamon Press, Addison-Wesley Publishing Co.
Lange, M. (2009). *Laws and lawmakers, science, metaphysics, and the laws of nature*. New York: Oxford University Press.
Lewis, D. (1973). *Counterfactuals*. Harvard University Press.
Lewis, D. (1994). Humean supervenience debugged. *Mind, 103*, 473–490.
Lotze, H. (1887). *Outlines of logic and of encyclopedia of philosophy* (G. T. Ladd, Ed. and Trans.). Boston: Ginn & Co.
Markosian, N., & Carroll, J. W. (2010). *An introduction to metaphysics*. Cambridge University Press.
Maxwell J. C. *On Ohm's Law*. Reprinted from the British Association Report, 1876, reprinted in The Scientific Papers of James Clerk Maxwell, ed. W. D. Niven, MA., FRS. Two volumes bound as one volume. Volume 2, pp. 533–537.
Maxwell, J. C. *Matter and motion*, with notes and appendices by Sir Joseph Larmor. Original edition 1877. Dover Publications, Inc.
McGee, V. Ramsey's dialetheism. In G. Priest, B. Armour-Garb, & J. C. Beall (Eds.), *The law of non-contradiction* (p. 2005). Oxford University Press.
Mellor, D. H., & Ramsey, F. P. (1990). *Philosophical papers*. Cambridge University Press.
Mill, J. S. (1843). *A system of logic*.
Morgenbesser, S., & Koslow, A. (2010). Theories and their worth. *The Journal of Philosophy, CVII, 12*, 615–647.
Nagel, E. (1961). *The structure of science*. Harcourt, Brace, & World, Inc.
Nagel, E. (1979). *Teleology revisited and other essays in the philosophy and history of science*. Columbia University Press.
Newman, M. H. A. Mr. Russell's Causal theory of perception. *Mind, 37*(146), 137–148.
Newton, I. (1947). *Sir Isaac Newton's mathematical principles of natural philosophy and his system of the world* (F. Cajori, Trans.). University of California Press.
Pippard, A. B. (1957). *The elements of classical thermodynamics*. Cambridge: Cambridge University Press.
Psillos, S. (2002). *Causation and explanation*. McGill- Queens University Press.
Ramsey, F. P. (1990). *Philosophical papers* (D. H. Mellor, Ed.). Cambridge University Press.
Roberts, J. T. (2008). *The law-governed universe*. Oxford University Press.
Ruben, D.-H. (1992). *Explaining explanation*. London: Routledge.
Russell, B., & Whitehead A. N. (1997). *Principia mathematica*, in PM to $*56$. Cambridge Mathematical Library Pb.
Sauer, T., & Majer, U (eds.). (2009). *Hilbert's lectures on the foundations of physics, 1915–1927*. Springer.
Sullivan, P. M. (1995). Wittgenstein on "The foundations of mathematics", June 1927. *Theoria, 61* (2), 105–142.
Suppes, P. (2002). *Representation and invariance of scientific structures*. Stanford: CSLI Publications.
Tao, T. (2008). Hamiltonians. In T. Gowers (Ed.) *The Princeton companion to mathematics*, J Barrow-Green and I Leader (associate eds.). Oxford: Princeton University Press.
Weyl, H. (1949). *Philosophy of mathematics and natural science*. Princeton University Press.
Whitney, H. (1968). The mathematics of physical quantities. Part I: Mathematical models for measurement. Part II: Quantity structures and dimensional analysis. *American Mathematical Monthly, 75*, 115–138, 226–256.

# Index



 
A. Koslow, *Laws and Explanations; Theories and Modal Possibilities*, Synthese Library 410, https://doi.org/10.1007/978-3-030-18846-7

Printed in the United States
by Baker & Taylor Publisher Services

Printed in the United States
by Baker & Taylor Publisher Services